I0511064

APPLIED RESEARCH IN
FIELD CROP PATHOLOGY
FOR INDIANA, 2024

APPLIED RESEARCH IN FIELD CROP PATHOLOGY FOR INDIANA, 2024

DARCY E. P. TELENKO
AND SUJOUNG SHIM

PURDUE UNIVERSITY PRESS
WEST LAFAYETTE, INDIANA

Copyright © 2025 by Darcy E. P. Telenko and Sujoung Shim. All rights reserved.

Cataloging-in-Publication Data on file at the Library of Congress.

ISBN 978-1-62671-236-2 (paperback)

ISBN 978-1-62671-201-0 (epdf)

CONTENTS

ACKNOWLEDGMENTS

This report is a summary of applied field crop pathology research trials conducted in 2024 under the direction of the Purdue Field Crop Pathology program in the Department of Botany and Plant Pathology at Purdue University. The authors wish to thank the Purdue Agronomy Research and Education Center, the Purdue Agricultural Centers, Purdue Extension Agriculture and Natural Resource educators, and the many cooperators and contributors who provided the resources needed to support the applied field crop pathology research program in Indiana.

Special recognition is extended to Sujoung Shim for technical skills in managing field trials and data organization and processing and for help in preparing this report; Steven Brand for trial establishment and management; Camila Rocca da Silva, Morgan Goodnight, Monica Mizuno, Ivis Miranda, Emily Duncan, Edward Peña Roncancio, Juan Peña Roncancio, and Mariana Moreano Acevedo, graduate students and visiting scholars who assisted with field trial data collection and analysis; Emilia Meyers, Emma Orr, Corbin Wentworth, and Destin Gentillon, undergraduate student interns who assisted with field trial data collection and scouting; Dr. Tom Creswell, Dr. John Bonkowski, and Tina Garwood with the Purdue Plant Pest Diagnostic Laboratory for assistance in pathogen surveys and diagnosis. Collectively, the contributions of colleagues, professionals, students, and growers were responsible for a highly successful and productive program to evaluate products and practices for disease management in field crops.

The authors would also like to thank the following for their support in 2024: Adama, Albaugh, Bayer Crop Science, BASF, Corteva Agriscience, FMC Agricultural Solution, Gowan, the Indiana Corn Marketing Council, the Indiana Soybean Alliance, the Koppert, National Predictive Modeling Tool Initiative, North Central Soybean Research Program, ProFarm, Maron Bio, Pioneer, Purdue University, Rovensa, Sipcam Agro, Syngenta, the USDA NIFA Hatch Project #IND00162952, the United Soybean Board, USWBSI-NFO, and Valent.

SUMMARY OF 2024 FIELD CROP DISEASE SEASON

CORN

In 2024, there was moderate disease on corn in Indiana across the state; details of major issues are listed below. Gray leaf spot, northern corn leaf blight, northern corn leaf spot, and southern rust were found in pockets. Tar spot and southern rust were two diseases that were closely monitored this season.

Tar spot. Tar spot of corn was a concern due to previous epidemics. In 2024, high levels of tar spot occurred in pockets across the state. The environmental conditions are key in determining field risk year to year, as temperature and leaf wetness plays an important role in tar spot disease development. The sixth year of tar spot–directed research has been completed in Indiana.

We continue to scout for tar spot across the state. Five new counties were confirmed with tar spot in 2024. This is the *first time all 92 counties were confirmed with active tar spot during a single season* (Figure 1). It is

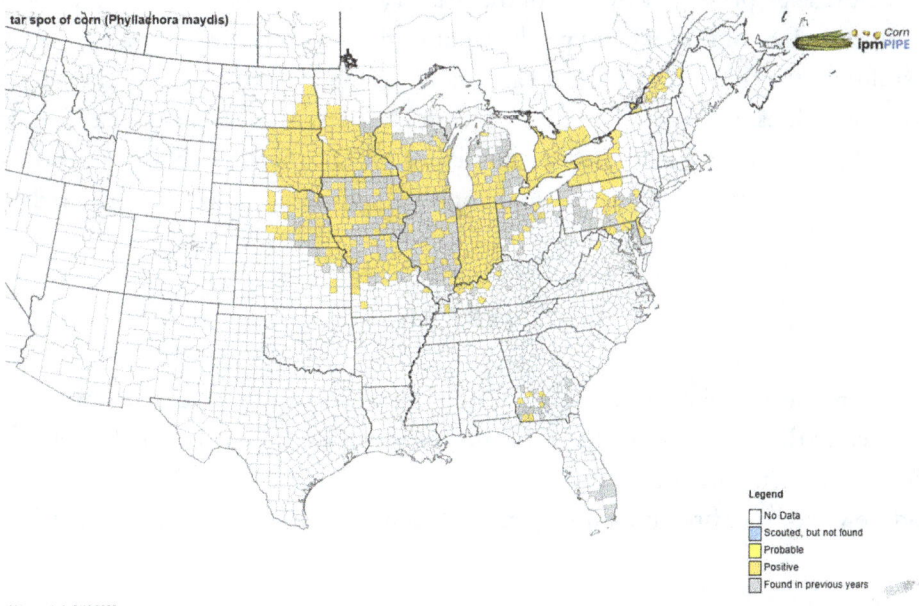

FIGURE 1. 2024 tar spot tracking across the United States and Canada. Yellow indicates that a positive sample was collected from that county during the 2024 season, and gray indicates that tar spot has been found previously. Image source: IPM Pipe, https://corn.ipmpipe.org/tarspot/.

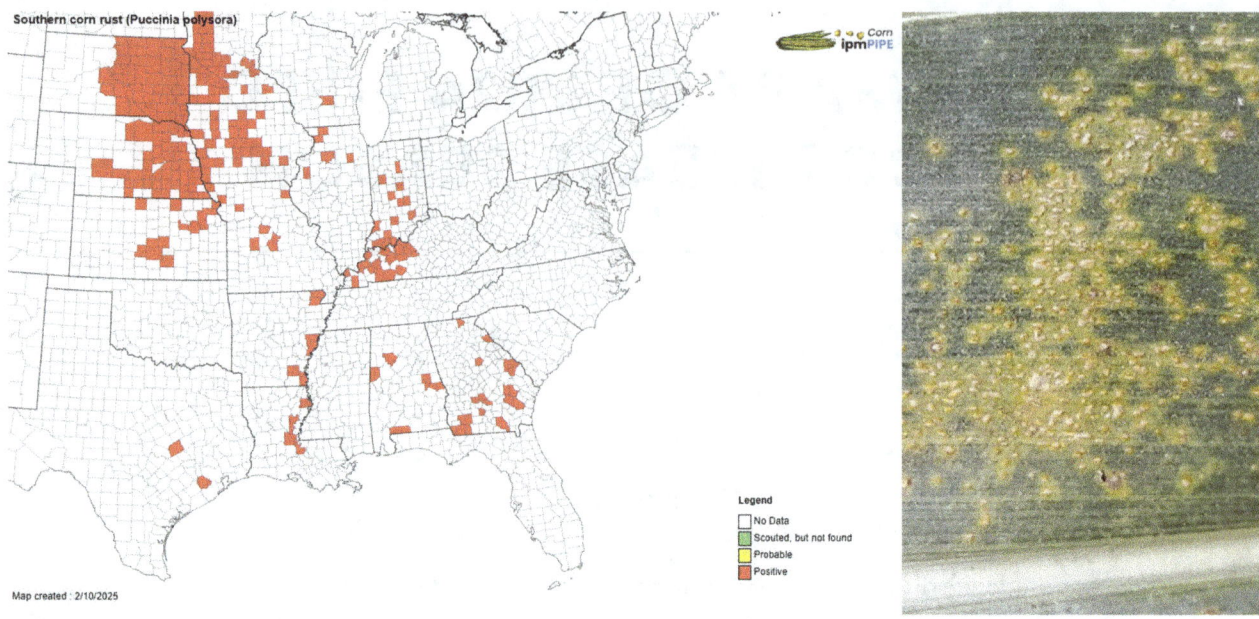

FIGURE 2. Southern corn rust map of confirmed (red) counties that had southern corn rust in 2024 and a corn leaf with southern rust infection. Photo credit: D. Telenko. Map source: IPM Pipe, https://corn.ipmpipe.org/southerncornrust/.

important to document tar spot movement in the state so that when favorable conditions arise, tar spot disease risk can be more accurately assessed across the remainder of the state.

Southern corn rust. Southern corn rust was first found in Indiana on July 25, 2024. A total of 14 counties were confirmed to have the disease present by the end of the season (Figure 2). Southern rust pustules generally tend to occur on the upper surface of the leaf and produce chlorotic symptoms on the underside of the leaf (Figure 2). These pustules rupture the leaf surface and are orange to tan in color. They are circular to oval in shape. Common rust was also widespread, and both diseases could be present on a leaf and easily mistaken for each other. It is important to send a sample to the Purdue Plant Pest Diagnostic Lab for confirmation if southern rust is suspected. There is an increased risk for yield impact if southern rust is identified early in the season.

SOYBEAN

Diseases in soybeans remained relatively low throughout the season for much of the state. Our research sites and sentinel plots across the state saw low levels of frogeye leaf spot, Cercospora leaf blight, Septoria brown spot, downy mildew, and white mold. There were also pockets where sudden death syndrome caused issues in fields. We had new reports of red crown rot, which has now been confirmed in eight Indiana counties (Figure 3).

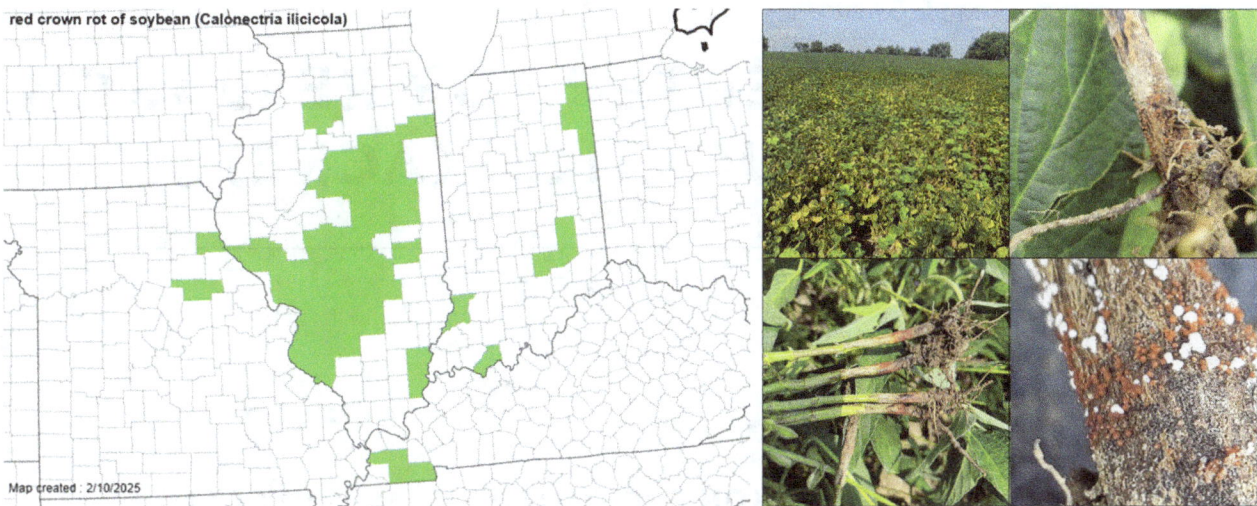

FIGURE 3. Red crown rot of soybean map and symptoms of red crown rot in the field, with typical red crown discoloration and orange-reddish spore-bearing structures that can form on the soybean crown/stem. Photo credit: D. Telenko. Map source: Crop Protection Network, https://cropprotectionnetwork.org/maps/red-crown-rot-map.

WHEAT

Fusarium head blight (FHB), or scab, is one of the most impactful diseases of wheat and among the most challenging to prevent. In addition, FHB infection can cause the production of a mycotoxin called deoxynivalenol (DON), or vomitoxin. The conditions in 2024 were moderately conducive to FHB development. Our research sites in both West Lafayette and Vincennes had moderate levels of FHB develop in our nontreated susceptible cultivar checks, and DON detection ranged anywhere from 4.0 to 9.4 ppm in the grain (Figure 4). FHB management requires an integrated approach, including the selection of cultivars with moderate resistance and timely fungicide application at flowering. Very few other diseases were observed in our wheat trials.

FIGURE 4. Wheat head exhibiting Fusarium head blight symptoms in a field. An example of the bleached spikelets on an infected head with pink/salmon fugal sporulation. Photo credit: D. Telenko.

AGRONOMY CENTER FOR RESEARCH AND EDUCATION (ACRE)

FUNGICIDE EVALUATION FOR FOLIAR DISEASES IN CORN IN CENTRAL INDIANA, 2024 (COR24-01.ACRE)

J. D. Peña, K. M. Goodnight, and D. E. P. Telenko, Department of Botany and Plant Pathology, Purdue University, West Lafayette, IN 47907-2054

CORN (*ZEA MAYS* P0574AM)

Tar spot, *Phyllachora maydis*
Gray leaf spot, *Cercospora zeae-maydis*
Northern corn leaf blight *Exserohilum turcicum*

A trial was established at the Purdue Agronomy Center for Research and Education (ACRE) in Tippecanoe County, Indiana. The experiment was a randomized complete block design with four replications. Plots were 10 feet wide and 30 feet long and consisted of four rows, and the two center rows were used for evaluation. The previous crop was corn. Standard practices for nonirrigated grain corn production in Indiana were followed. Corn hybrid P0574AM was planted in 30-inch row spacing at a rate of 34,000 seeds/acre on May 13. Foliar applications were made at tassel/silk (VT/R1) growth stage on July 20. All foliar fungicide applications were applied at 15 gal/acre and at 40 psi using a Lee self-propelled sprayer equipped with a 10-foot boom, fitted with six TJ-VS 8002 nozzles spaced 20 inches apart. Disease ratings were assessed on September 3 at dent (R5) growth stage. Tar spot, gray leaf spot (GLS), and northern corn leaf blight (NCLB) severity was visually assessed as a percentage (0–100%) of affected leaf area at ear leaf on five plants in each plot, and values were averaged before analysis. Lodging was evaluated on September 16 by determining the percentage of lodged stalks when pushed from shoulder height to 45° from vertical. The two center rows of each plot were harvested on October 8, and yields were adjusted to 15.5% moisture. All disease and yield data were analyzed in SAS 9.4 (SAS Institute, Cary, NC). A generalized linear mixed model analysis of variance was performed using PROC GLIMMIX. Values are least squares means, and values with different letters are significantly different based on Fisher's least significant difference (α = 0.05).

In 2024, weather conditions were moderately favorable for disease development. Tar spot, GLS, NCLB, and southern rust were present in the trial. There was no significant effect of fungicide treatments on reducing foliar disease severity over nontreated control (Table 1). No significant differences were observed for canopy greenness, lodging, harvest moisture, test weight, and yield of corn.

TABLE 1. *Effect of Treatments on Foliar Disease Severity, Canopy Greenness, Lodging, and Yield of Corn*

TREATMENT AND RATE/ACRE[z]	TOTAL LEAF DISEASE[y] %	CANOPY GREEN[x] %	LODGING[w] %	HARVEST MOISTURE %	TEST WEIGHT LB/BU	YIELD[v] BU/ACRE
Nontreated control	4.8	25.0	7.8	17.1	57.5	208.8
Veltyma 3.34 SC 7.0 fl oz	2.7	31.3	0.0	17.5	59.2	217.0
Delaro Complete 3.82 SC 8.0 fl oz	1.1	30.8	5.0	17.3	55.1	214.5
Aproach Prima 2.34 SC 6.8 fl oz	1.7	27.5	2.5	17.7	55.7	210.7
Adastrio 4.0 SC 8.0 fl oz	1.6	27.5	7.5	17.6	55.3	204.2
Miravis Neo 2.5 SE 13.7 fl oz	1.1	30.0	2.5	17.7	55.4	201.0
Trivapro 2.21 SE 13.7 fl oz	1.0	31.3	2.5	17.6	55.4	211.8
Headline AMP 1.68 SC 10.0 fl oz	1.9	28.3	2.5	17.5	55.6	212.7
Proline 4.0 SC 5.7 fl oz	4.0	27.5	5.0	17.6	55.3	208.6
Quadris 2.08 SC 6.0 fl oz	1.9	28.3	5.0	17.4	56.7	208.5
Tilt 3.6 ES 4.0 fl oz	3.0	28.8	5.0	17.4	55.3	210.8
P-value[u]	*0.0682*	*0.4710*	*0.8799*	*0.9313*	*0.5933*	*0.5839*

[z] Foliar applications were made at tassel/silk (VT/R1) growth stage on July 20. All foliar fungicide applications were applied at 15 gal/acre.

[y] Foliar disease severity (tar spot, gray leaf spot, northern corn leaf blight and southern rust) visually assessed as a percentage (0–100%) of symptomatic leaf area on ear leaf on September 3. Five plants were assessed per plot and averaged before analysis.

[x] Canopy greenness as %, visually rated per plot was assessed on September 16 at dent (R5).

[w] Lodging was evaluated on September 25 by determining the percentage of lodged stalks when pushed from shoulder height to the 45° from vertical.

[v] Yields were adjusted to 15.5% moisture and harvested on October 8.

[u] All data were analyzed in SAS 9.4 (SAS Institute, Cary, NC). A generalized linear mixed model analysis of variance was performed using PROC GLIMMIX. Values are least squares means, and values with different letters are significantly different based on Fisher's least significant difference ($\alpha = 0.05$).

EVALUATION OF TILLAGE, HYBRID, AND FUNGICIDE FOR FOLIAR DISEASES IN CORN IN CENTRAL INDIANA, 2024 (COR24-04.ACRE)

M. Acevedo, S. Shim, and D. E. P. Telenko, Department of Botany and Plant Pathology, Purdue University, West Lafayette, IN 47907-2054

CORN (ZEA MAYS W2585VT2PRIB AND P0589AMXT)

Tar spot, *Phyllachora maydis*
Gray leaf spot, *Cercospora zeae-maydis*
Northern corn leaf blight, *Exserohilum turcicum*

A trial was established at the Purdue Agronomy Center for Research and Education (ACRE) in Tippecanoe County, Indiana. The experiment was a split plot with six replications. Plots were 10 feet wide and 30 feet long and consisted of four rows, and the two center rows were used for evaluation. The previous crops were no-till corn and full-till soybean. Standard practices for nonirrigated grain corn production in Indiana were followed. The tillage blocks (no-till and full-tillage) were the main effect. Two corn hybrids and a fungicide application (yes/no) were factorial arrangements in the subplots. Corn hybrids W2585VT2PRIB and P0589AMXT were planted in 30-inch row spacing at a rate of 2 seeds/foot on May 14. Veltyma fungicide at 7.0 fl oz/acre was applied at the tasseling/silking (VT/R1) growth stage. All foliar fungicide applications were applied at 15 gal/acre and 40 psi using a Lee self-propelled sprayer equipped with a 10-foot boom, fitted with six TJ-VS 8002 nozzles spaced 20 inches apart. Disease ratings were assessed on September 17 at the dent/maturity (R5/R6) growth stage and were rated by visually assessing the percentage (0–100%) of symptomatic leaf area at ear leaf on 10 plants in each plot. Values for each plot were averaged before analysis. The two center rows of each plot were harvested on October 8, and yields were adjusted to 15.5% moisture. All data were analyzed in SAS 9.4 (SAS Institute, Cary, NC). A generalized linear mixed model analysis of variance was performed using PROC GLIMMIX. Values are least squares means, and values with different letters are significantly different based on Fisher's least significant difference (α = 0.05).

In 2024, weather conditions were favorable for disease. Tar spot, gray leaf spot (GLS), and northern corn leaf blight (NCLB) were present in the trial. There was a significant interaction found with tillage, hybrid, and fungicide treatments for disease severity; therefore, simple effects were analyzed. Tar spot severity was more severe in the full-tillage plots planted to W2585VT2RIB and not treated with a fungicide, followed by no-till plots planted to W2585VT2RIB and not treated by a fungicide (Table 2). The P0589AMXT hybrid had significantly less tar spot than W2585VT2RIB when nontreated with a fungicide. A fungicide application at VT/R1 significantly reduced disease in the susceptible hybrid under both tillage treatments. A fungicide application also significantly reduced disease in the moderately resistant hybrid when planted in full tillage. GLS had higher severity in the no-till plot planted to W2585VT2RIB and not treated by a fungicide. The addition of a fungicide at VT/R1 significantly reduced GLS in the no-till plots in both hybrids. There were no significant differences between treatments under full tillage, as all had a low level of GLS. NCLB had the highest severity in no-till plots with both hybrids not treated with a fungicide. A fungicide application significantly reduced NCLB in the no-till plots; there were no significant differences between hybrids and treatment in the full-tillage plots. There was no significant difference in treatments for lodging. Harvest moisture was highest in W2585VT2RIB plus fungicide in both tillage treatments. Test weight was highest under full tillage, with Po589AMXT and not treated with a fungicide. Corn yield was highest in the full-tillage plots with hybrid W2585VT2PRIB with a fungicide application, but this was not significantly different from W2585VT2PRIB nontreated or Po589AMXT treated with a fungicide.

TABLE 2. *Effect of Tillage, Hybrid and Fungicide on Foliar Disease Risk in Corn and Yield of Corn*

TILLAGE, HYBRID, TREATMENT, AND RATE/A[z]	TAR SPOT %[y]	GLS %[y]	NCLB %[y]	LODGING %	HARVEST MOISTURE %	TEST WEIGHT LB/BU	YIELD[x] BU/ACRE
No till, W2585VT2PRIB, Nontreated control	17.6 b	7.2 a	19.5 a	3.3	17.7 b	55.0 c	208.0 d
No till, W2585VT2PRIB, Veltyma 7.0 fl oz	0.4 d	1.1 cd	5.9 bc	3.3	18.7 a	54.9 c	219.9 c
No till, Po589AMXT, Nontreated control	1.0 d	5.1 b	13.9 a	10.0	16.8 d	56.2 ab	199.3 d
No till, Po589AMXT, Veltyma 7.0 fl oz	0.3 d	2.3 c	6.2 bc	5.0	17.6 b	56.3 ab	205.0 d
Full till, W2585VT2PRIB, Nontreated control	22.5 a	0.9 d	4.7 bc	6.7	17.5 bc	56.0 ab	254.4 ab
Full till, W2585VT2PRIB, Veltyma 7.0 fl oz	7.1 c	0.1 d	0.3 c	1.7	18.5 a	55.6 bc	263.6 a
Full till, Po589AMXT, Nontreated control	5.6 c	0.9 d	7.3 b	1.7	16.3 c	57.0 a	251.1 b
Full till, Po589AMXT, Veltyma 7.0 fl oz	2.3 d	0.1 d	0.5 c	0.0	16.8 cd	55.7 bc	257.1 ab
P-value[w]	0.0001	0.0001	0.0001	0.0982	0.0001	0.0044	0.0001

[z] The fungicide (Veltyma 7.0 fl oz/acre) was applied at 15 gal/acre and 40 psi using a Lee self-propelled sprayer equipped with a 10-foot boom, fitted with six TJ-VS 8002 nozzles spaced 20 inches apart and applied on August 2 at tasseling/silking (VT/R1) growth stage.

[y] Foliar disease severity visually assessed as a percentage (0–100%) of symptomatic leaf area on ear leaf, with 10 plants assessed per plot and averaged before analysis on September 17 at dent/maturity (R5/R6) for tar spot stromata and on September 10 at dent/maturity (R5/R6) for GLS and NCLB. GLS = gray leaf spot; NCLB = northern corn leaf blight.

[x] Yields were adjusted to 15.5% moisture and harvested on October 8.

[w] All data were analyzed in SAS 9.4 (SAS Institute, Cary, NC). A generalized linear mixed model analysis of variance was performed using PROC GLIMMIX. Values are least squares means, and values with different letters are significantly different based on Fisher's least significant difference ($\alpha = 0.05$).

EVALUATION OF SEED TREATMENTS FOR PYTHIUM IN CORN IN CENTRAL INDIANA, 2024 (COR24-32.ACRE)

S. Shim and D. E. P. Telenko, Department of Botany and Plant Pathology, Purdue University, West Lafayette, IN 47907-2054

CORN (ZEA MAY)

Seeding disease, *Pythium sylvaticum*

A trial was established at the Purdue Agronomy Center for Research and Education (ACRE) in Tippecanoe County, Indiana. The experiment was a randomized complete block design with four replications. Plots were 10 feet wide and 30 feet long and consisted of four rows, and the two center rows were used for evaluation. The previous crop was corn. Standard practices for nonirrigated grain corn production in Indiana were followed. Corn was planted in 30-inch row spacing at a rate of 2 seeds/foot on May 13. Treated seeds were provided by the cooperator. Inoculum of *Pythium sylvaticum* was applied within the seedbed at 1.24 g/foot at planting. Stand counts were assessed on June 17 at V4 growth stage. The two central rows of each plot were harvested on October 9. All data were analyzed in SAS 9.4 (SAS Institute, Cary, NC). A generalized linear mixed model analysis of variance was performed using PROC GLIMMIX. Values are least squares means, and values with different letters are significantly different based on Fisher's least significant difference (α = 0.05).

In 2024, weather conditions were moderately favorable for disease development. The seed treatment of Thiamethoxam 2.5 fl oz + M-5P23 1.0 fl oz had significantly less stand than control of Thiamethoxam 2.5 fl oz and Thiamethoxam 1.28 fl oz + M-123 EP 0.8 fl oz + M-5P23 0.5 fl oz on Jun 17. All seed treatments increase canopy greenness at dent (R5) over the control of Thiamethoxam 2.5 fl oz (Table 3). There was no significant effect of seed treatment on harvest moisture, test weight, and yield of corn.

TABLE 3. *Effect of Seed Treatment on Stand Count, Canopy Greenness, and Yield of Corn*

SEED TREATMENT AND RATE/CWT[z]	STAND COUNT[y] #/ACRE	CANOPY GREEN[x] %	HARVEST MOISTURE %	TEST WEIGHT LB/BU	YIELD[w] BU/ACRE
Control: Thiamethoxam 2.5 fl oz	36,881 a	76.3 c	15.6	62.0	211.4
Thiamethoxam 2.5 fl oz + mefenoxam 0.08 fl oz	35,647 ab	81.3 b	15.4	61.4	202.9
Thiamethoxam 2.5 fl oz + M-123 EP 0.8 fl oz	35,864 ab	82.5 ab	15.4	61.3	200.6
Thiamethoxam 2.5 fl oz + M-5P23 0.5 fl oz	35,574 ab	81.3 b	15.5	61.3	194.9
Thiamethoxam 2.5 fl oz + M-5P23 1.0 fl oz	34,267 b	85.0 a	15.8	60.4	200.5
Thiamethoxam 1.28 fl oz + M-123 EP 0.8 fl oz + M-5P23 0.5 fl oz	37,462 a	83.8 ab	13.1	56.0	199.6
P-value[v]	0.0472	0.0012	0.1343	0.4213	0.6114

[z] All plots were inoculated with *Pythium sylvaticum* at 2.0 g/foot within the seedbed at planting on May 13.

[y] Stand count was assessed on June 17 at V4 growth stage.

[x] Canopy greenness as % (0–100), visually rated per plot and assessed on September 5 at R5.

[w] Yields were adjusted to 15.5% moisture and harvested on October 9.

[v] All data were analyzed in SAS 9.4 (SAS Institute, Cary, NC). A generalized linear mixed-model analysis of variance was performed using PROC GLIMMIX. Values are least squares means, and values with different letters are significantly different based on Fisher's least significant difference (α = 0.05).

FUNGICIDE COMPARISON FOR FOLIAR DISEASES ON SHORT CORN IN CENTRAL INDIANA, 2024 (COR24-38.ACRE)

E. Schillinger, S. Shim, and D. E. P. Telenko, Department of Botany and Plant Pathology, Purdue University, West Lafayette, IN 47907-2054

CORN (*ZEA MAYS* PR111-20SSC AND PR108-20SSC)

Tar spot, *Phyllachora maydis*
Gray leaf spot, *Cercospora zeae-maydis*
Northern corn leaf blight, *Setosphaeria turcica*

A trial was established at the Purdue Agronomy Center for Research and Education (ACRE) in Tippecanoe County, Indiana. The experiment was a randomized complete block design with four replications. Plots were 10 feet wide and 30 feet long and consisted of four rows, and the two center rows were used for evaluation. The previous crop was corn. Standard practices for grain corn production in Indiana were followed. Corn hybrids PR111-20SSC and PR108-20SSC were planted in 30-inch row spacing at a rate of two seeds/foot on May 14. All fungicides were applied at 15 gal/acre and 40 psi using a Lee self-propelled sprayer equipped with a 10-foot boom fitted with six TJ-VS 8002 nozzles spaced 20 inches apart. Fungicides were applied on July 20 at tassel/silk (VT/R1) growth stage. Disease ratings were assessed on September 3 at dent (R5) growth stage. Tar spot, gray leaf spot (GLS), and northern corn leaf blight (NCLB) severity was visually assessed as a percentage (0–100%) of symptomatic leaf area at ear leaf on five plants per plot and averaged before analysis. The two center rows of each plot were harvested on October 8, and yields were adjusted to 15.5% moisture. All data were analyzed in SAS 9.4 (SAS Institute, Cary, NC). A generalized linear mixed model analysis of variance was performed using PROC GLIMMIX. Values are least squares means, and values with different letters are significantly different based on Fisher's least significant difference ($\alpha = 0.05$).

In 2024, weather conditions were moderately favorable for foliar diseases. Tar spot, GLS, and NCLB were the most prominent diseases in the trial. There was no significant interaction between hybrid and fungicide application; therefore, main effects are shown (Table 4). There was no significant difference between hybrids for foliar disease severity except for PR111-20SSC, which had significantly less NCLB compared to PR108-20SSC. PR111-20SSC showed significantly increased grain moisture and yield over PR108-20SSC, but there were no significant differences in test weight between hybrids. Delaro Complete significantly reduced tar spot, GLS, and NCLB severity over the nontreated control. There were no significant differences between treatments for harvest moisture, test weight, and yield of corn.

TABLE 4. *Effect of Treatment on Foliar Diseases and Yield of Corn*

TREATMENT[z]	TAR SPOT[y] % STROMATA	GLS[y] % SEVERITY	NCLB[y] % SEVERITY	HARVEST MOISTURE %	TEST WEIGHT LB/BU	YIELD[x] BU/ACRE
Hybrid						
PR111-20SSC	1.1	1.0	0.6 b	21.9 a	53.3	207.2 a
PR108-20SSC	0.9	0.8	1.7 a	19.7 b	54.9	191.8 b
Fungicide						
Nontreated control	1.7 a	1.2 a	1.7 a	21.1	54.8	201.6
Delaro Complete 3.82 SC 8.0 fl oz/acre	0.4 b	0.5 b	0.7 b	20.4	53.5	197.3
P-value hybrid[w]	0.5077	0.2793	0.0106	0.0009	0.4611	0.0040
P-value fungicide	0.0005	0.0032	0.0301	0.2125	0.5890	0.2405
P-value hybrid*fungicide	0.8459	0.7464	0.3046	0.6046	0.6867	0.1026

[z] Fungicides were applied on July 20 at tassel/silk (VT/R1) growth stage.

[y] Foliar disease was visually assessed as a percentage (0–100%) of affected leaf area on five plants in each plot at the ear leaf on at dent (R5) growth stage on September 3. GLS = gray leaf spot; NCLB = northern corn leaf blight.

[x] Yields were adjusted to 15.5% moisture and harvested on October 8.

[w] All data were analyzed in SAS 9.4 (SAS Institute, Cary, NC). A generalized linear mixed model analysis of variance was performed using PROC GLIMMIX. Values are least squares means, and values with different letters are significantly different based on Fisher's least significant difference (α = 0.05).

COMPARISON OF FUNGICIDE EFFICACY FOR FOLIAR DISEASES OF SOYBEAN IN CENTRAL INDIANA, 2024 (SOY24-01.ACRE)

K. M. Goodnight, S. Shim, and D. E. P. Telenko, Department of Botany and Plant Pathology, Purdue University, West Lafayette, IN 47907-2054

SOYBEAN (*GLYCINE MAX* P29A19E)

Frogeye leaf spot, *Cercospora sojina*
Septoria brown spot, *Septoria glycines*
Cercospora leaf blight, *Cercospora* spp.

A trial was conducted at the Purdue Agronomy Center for Research and Education (ACRE) in Tippecanoe County, Indiana. The experiment was a randomized complete block design with four replications. Plots were 10 feet wide and 30 feet long and consisted of four rows, and the two center rows were utilized for evaluation. The previous crop was corn. Standard practices for soybean production in Indiana were followed. Soybean cultivar P29A19E was planted in 30-inch row spacing at a rate of 140,000 seeds/acre on May 8. Fungicide applications were applied on July 18 at beginning pod (R3) and were applied at 15 gal/acre at 40 psi using a Lee self-propelled sprayer equipped with a 10-foot boom, fitted with six TJ-VS 8002 nozzles spaced 20 inches apart, at 3.6 mph. Disease ratings were assessed on August 31 at full seed growth stage (R6). Frogeye leaf spot was rated for disease severity by visually assessing the symptomatic leaf area in the canopy. Canopy defoliation was visually rated on a scale of 0–100% on August 31. Percent green stem was visually assessed as a percentage (0–100%) of plants in the plot on September 26. The two center rows of each plot were harvested on September 26, and yields were adjusted to 13% moisture. All data were analyzed in SAS 9.4 (SAS Institute, Cary, NC). A generalized linear mixed model analysis of variance was performed using PROC GLIMMIX. Values are least squares means, and values with different letters are significantly different based on Fisher's least significant difference ($\alpha = 0.05$).

In 2024, weather conditions were not favorable for disease development. Frogeye leaf spot, Septoria brown spot, and Cercospora leaf blight were the most prominent diseases in the trial and reached low severity. All fungicide treatments significantly reduced frogeye leaf spot severity compared to the nontreated control (Table 5). Veltyma, Echo + Folicur + Topsin, Delaro Complete, Topguard, and Revyteck significantly reduced canopy defoliation as compared to the nontreated control on August 31. There was significantly more green stem with a treatment of Trivapro and Veltyma over nontreated control at harvest. There were no differences between fungicides and the nontreated control for yield of soybean.

TABLE 5. *Effect of Treatments on Foliar Diseases and Yield of Soybean*

TREATMENT AND RATE/ACRE[z]	FLS % SEVERITY[y]	DEFOLIATION %[x]	GREEN STEM %[w]	YIELD[v] BU/ACRE
Nontreated control	0.7 a	5.0 a	2.5 cd	78.6
Topguard EQ 4.29 SC 5.0 fl oz	0.1 b	3.0 bc	3.5 bcd	80.1
Lucento 4.17 SC 5.0 fl oz	0.2 b	3.8 ab	3.5 bcd	78.9
Trivapro 13.7 fl oz	0.1 b	4.3 ab	6.5 ab	76.8
Quadris 2.08 SC 6.0 fl oz	0.3 b	4.3 ab	2.0 d	77.0
Veltyma 3.34 SC 7.0 fl oz	0.2 b	1.5 c	10.0 a	77.7
Revytek 3.33 SC 8.0 fl oz	0.1 b	2.5 bc	4.0 bcd	79.3
Echo 2.21 SE 36.0 fl oz + Folicur 4.0 fl oz + Topsin 4.5 SC 20.0 fl oz	0.1 b	1.5 c	4.8 bcd	78.5
Delaro Complete 3.83 SC 8.0 fl oz	0.2 b	1.5 c	5.8 bc	79.3
Miravis Neo 2.5 EC 13.7 fl oz	0.2 b	4.0 ab	5.3 bcd	78.7
Topsin 4.5 SC 20.0 fl oz	0.3 b	4.0 ab	4.0 bcd	78.3
P-value[u]	0.0338	0.0042	0.0089	0.8042

[z] Fungicide applications were made on July 18 at full seed (R3) growth stage and contained a nonionic surfactant (Preference) at a rate of 0.25% v/v.

[y] Foliar disease severity rated on a scale of 0–100% of canopy within a plot with disease symptoms on August 31 at full seed growth stage (R6). FLS = frogeye leaf spot.

[x] Defoliation was visually rated on a scale of 0–100% on August 31.

[w] Percent green stem was visually assessed as a percentage (0–100%) of plants in plot on September 26.

[v] Yields were adjusted to 13% moisture and harvested on September 26.

[u] All data were analyzed in SAS 9.4 (SAS Institute, Cary, NC). A generalized linear mixed model analysis of variance was performed using PROC GLIMMIX. Values are least squares means, and values with different letters are significantly different based on Fisher's least significant difference (α = 0.05).

EVALUATION OF SEED TREATMENT FOR SUDDEN DEATH SYNDROME ON SOYBEAN IN CENTRAL INDIANA, 2024 (SOY24-03.ACRE)

E. Myers, S. Shim, and D. E. P. Telenko, Department of Botany and Plant Pathology, Purdue University, West Lafayette, IN 47907-2054

SOYBEAN (*GLYCINE MAX* AG26XF1)

Sudden death syndrome, *Fusarium virguliforme*

A trial was established at the Purdue Agronomy Center for Research and Education (ACRE) in Tippecanoe County, Indiana. The experiment was a randomized complete block design with four replications. Plots were 10 feet wide and 30 feet long and consisted of four rows, and the two center rows were used for evaluation. The previous crop was corn. Standard practices for soybean production in Indiana were followed. Soybean cultivar AG26XF1 was planted in 30-inch row spacing at a rate of 8 seeds/foot on April 18. *Fusarium virguliforme* inoculum was applied at planting at 1.25 g/foot within the seedbed. Seed treatments were applied on seeds before planting. In-furrow and 2x2 applications were applied at planting at 10 gal/acre. Fungicide applications were applied on July 2 at full bloom (R2) and were applied at 15 gal/acre at 40 psi using a Lee self-propelled sprayer equipped with a 10-foot boom, fitted with six TJ-VS 8002 nozzles spaced 20 inches apart, at 3.6 mph. Disease ratings were assessed on August 30 at beginning maturity (R7) growth stage. Sudden death syndrome (SDS) in each plot was rated for disease incidence (DI) as percentage of plants with disease symptoms (0–100%) and disease severity (DS) on a scale of 1 to 9, where 1 refers to low disease pressure and 9 refers to premature death of the plant. The SDS index (DX) was calculated using the equation $DX = (DI*DS)/9$. Ten roots per plot were sampled from border rows at the R4 (full pod) growth stage on August 9, gently washed, and rated for root rot severity on a scale of 0–100%. The two center rows of each plot were harvested on September 26, and yields were adjusted to 13% moisture. All data were analyzed in SAS 9.4 (SAS Institute, Cary, NC). A generalized linear mixed model analysis of variance was performed using PROC GLIMMIX. Values are least squares means, and values with different letters are significantly different based on Fisher's least significant difference ($\alpha = 0.05$).

In 2024, weather conditions were not favorable for disease development, and very little disease developed in plots. SDS was present in the trial and reached low severity. There were no significant differences between nontreated control, base, and other seed treatments for the SDS index (Table 6). All treatments reduced root rot severity compared to the nontreated control. There was no significant effect of treatment on harvest moisture, test weight, and yield of soybean.

TABLE 6. *Valuation of Seed Treatment on Sudden Death Syndrome, Root Rot, and Yield of Soybean*

TREATMENT, RATE/ACRE, AND TIMING[z]	SDS INDEX[y]	ROOT ROT % SEVERITY[x]	HARVEST MOISTURE %	TEST WEIGHT LB/BU	YIELD[w] BU/ACRE
Nontreated control	13.2	5.0 a	15.8	52.7	68.1
Base	6.7	1.2 b	15.6	52.3	72.6
IleVO	2.6	1.3 b	15.9	53.3	71.1
Saltro	3.6	0.9 b	15.6	53.5	68.4
Zeltera	8.6	0.9 b	15.6	53.5	68.5
Base + Xylem Plus in-furrow 32.0 fl oz fb Xylem Plus 24.0 fl oz at R2	11.7	1.2 b	16.1	52.8	67.7
Base + Xyway 15.2 fl oz in 2x2	9.9	1.4 b	15.6	53.6	67.2
Base + ILeVO + Ceramax	4.4	0.8 b	16.2	52.9	72.3
P-value[v]	0.2115	0.0053	0.5813	0.6550	0.7201

[z] Seed treatments were applied on seeds before planting. Base contained Allegiance Fl at 4.0 g ai/100 kg + Stamina at 7.5 g ai/100 kg + Systiva XS Xemium Brand at 5.0 g ai/100 kg + Poncho 600 at 0.11 mg ai/seed. In-furrow and 2x2 applications were applied at planting at 10 gal/acre. Xylem plus was applied on July 2 at the full bloom (R2) growth stage. fb = followed by.

[y] Sudden death syndrome (SDS) index (DX) was calculated using the equation $DX = (DI*DS/9)$.

[x] Root rot was visually assessed as a percentage (0–100%) of dark discoloration on roots on August 9.

[w] Yields were adjusted to 13% moisture and harvested on September 26.

[v] All data were analyzed in SAS 9.4 (SAS Institute, Cary, NC). A generalized linear mixed model analysis of variance was performed using PROC GLIMMIX. Values are least squares means, and values with different letters are significantly different based on Fisher's least significant difference ($\alpha = 0.05$).

COMPARISON OF PLANTING DATES AND SEED TREATMENTS ON SOYBEAN IN CENTRAL INDIANA, 2024 (SOY24-08.ACRE)

I. L. Miranda, S. Shim, and D. E. P. Telenko, Department of Botany and Plant Pathology, Purdue University, West Lafayette, IN 47907-2054

SOYBEAN (*GLYCINE MAX* 24E453N)

Septoria brown spot, *Septoria glycines*

A trial was established at the Purdue Agronomy Center for Research and Education (ACRE) in Tippecanoe County, Indiana. The experiment design was a split plot with four replications. The main plot was planting date and subplot seed treatments. Plots were 10 feet wide and 30 feet long and consisted of four rows, and the two center rows were used for evaluation. The previous crop was corn. Standard practices for soybean production in Indiana were followed. Soybean cultivar 24E453N were planted in 30-inch row spacing at a rate of 8 seeds/foot. Treatments were a factorial arrangement of four planting dates by four seed treatments. Soybeans were planted on April 18 (planting date 1), May 2 (planting date 2), May 20 (planting date 3), and May 30 (planting date 4). Stand counts were assessed at cotyledons expanded/first-node (VC/V1) growth stage for each planting date. Disease ratings were assessed on August 31 at full seed/beginning maturity (R6/R7) growth stages. Septoria brown spot (SBS) was rated for disease severity by visually assessing the percentage of symptomatic leaf area in the upper and lower canopies. Canopy greenness was visually assessed on a scale of 0–100% green within a plot on August 31. Ten roots were sampled for outer rows of each plot on September 11 at the beginning maturity/full maturity (R7/R8) growth stages and were rated for root rot severity on a scale of 0–100% and averaged before analysis. The two center rows of each plot were harvested on October 3, and yields were adjusted to 13% moisture. All data were analyzed in SAS 9.4 (SAS Institute, Cary, NC). A generalized linear mixed model analysis of variance was performed using PROC GLIMMIX. Values are least squares means, and values with different letters are significantly different based on Fisher's least significant difference (α = 0.05).

In 2024, weather conditions were not favorable for disease development, and very little to moderate disease developed in plots. No significant interactions were observed between seed treatments on root rot; therefore, main effects of planting date and seed treatment were compared (Table 7). Soybean stand was reduced in the April 18 planting compared to the plantings on May 2, May 20, and May 30. The seed treatment CruiserMaxx APX + Thiamethoxam and CruiserMaxx APX had significantly higher stand than the nontreated control and Thiamethoxam. SBS severity was higher at planting on April 18 and May 2 compared to May 20 and May 30. No significant differences were detected between seed treatments on SBS and canopy greenness. The percentage of green had the highest value in the planting on May 30 compared to all the earlier planting dates. Root rot severity was significantly lower at the planting on May 20 and May 30 compared to the first planting on April 18, which had the highest root rot severity. There were no significant differences between planting dates and seed treatments on harvest moisture and test weight. Soybean yield was significantly higher at the first planting date (April 18) compared to later plantings. No significant differences were detected between seed treatments for yield of soybean.

TABLE 7. *Effect of Planting Dates and Seed Treatments on Stand Count, SBS, Canopy Greenness, Root Rot, and Yield of Soybean*

PLANTING DATES AND SEED TREATMENT[z]	STAND COUNT #/ACRE	SBS[y] %	CANOPY[x] % GREEN	ROOT ROT[w] %	HARVEST MOISTURE %	TEST WEIGHT LB/BU	YIELD[v] BU/ACRE
Planting Date							
Planting date 1 (April 18)	61,801 b	4.3 a	84.1 d	7.7 a	11.4	55.0	57.6 a
Planting date 2 (May 2)	81,457 a	3.9 a	87.8 c	4.0 b	11.4	55.2	49.4 b
Planting date 3 (May 20)	89,080 a	1.8 b	96.9 b	2.0 c	10.8	52.0	47.6 b
Planting date 4 (May 30)	92,565 a	1.0 c	100.0 a	0.7 c	11.2	55.2	46.3 b
Seed Treatment							
Nontreated control	69,533 b	2.9	93.2	4.6	10.6	51.6	49.3
CruiserMaxx APX + Thiamethoxam	89,570 a	2.5	91.4	3.7	11.4	55.2	50.0
Thiamethoxam	71,221 b	2.9	91.9	3.0	11.4	55.3	51.0
CruiserMaxx APX, no Thiamethoxam	94,580 a	2.8	92.2	3.2	11.4	55.4	50.5
P-value planting date[u]	0.0019	0.0001	0.0001	0.0001	0.5703	0.4756	0.0030
P-value seed treatment	0.0045	0.7108	0.5766	0.2672	0.3140	0.3488	0.9591
P-value planting date*seed treatment	0.2587	0.7905	0.6990	0.9180	0.3343	0.4108	0.9385

[z] Seed treatments applied prior to planting at 10 g AI/100 kg seed.

[y] Foliar disease severity rated on a scale of 0–100% of canopy within a plot with disease symptoms on August 31. SBS = Septoria brown spot.

[x] Canopy greenness was visually assessed on a scale of 0–100% green within a plot on August 31.

[w] Root rot was visually assessed as a percentage (0–100%) of dark discoloration on 10 roots per plot and was then averaged on September 11 at the beginning maturity/full maturity (R7/R8) growth stages.

[v] Yields were adjusted to 13% moisture and harvested on October 3.

[u] All data were analyzed in SAS 9.4 (SAS Institute, Cary, NC). A generalized linear mixed model analysis of variance was performed using PROC GLIMMIX. Values are least squares means, and values with different letters are significantly different based on Fisher's least significant difference (α = 0.05).

FROGEYE LEAF SPOT MODEL EVALUATION FOR FUNGICIDE APPLICATION IN SOYBEAN IN CENTRAL INDIANA, 2024 (SOY24-12.ACRE)

E. Peña, S. Shim, and D. E. P. Telenko, Department of Botany and Plant Pathology, Purdue University, West Lafayette, IN 47907-2054

SOYBEAN (*GLYCINE MAX* P29A19E)

Septoria brown spot, *Septoria glycines*
Frogeye leaf spot, *Cercospora sojina*

A trial was established at the Purdue Agronomy Center for Research and Education (ACRE) in Tippecanoe County, Indiana. The experiment was a randomized complete block design with four replications. Plots were 10 feet wide and 30 feet long and consisted of four rows, and the two center rows were used for evaluation. The previous crop was soybean. Standard practices for soybean production in Indiana were followed. Soybean cultivar P29A19E was planted in 30-inch row spacing at a rate of 140,000 seeds/acre on May 8. Fungicide applications were applied on July 18 at beginning pod (R3) and 30% model threshold and on August 1 at beginning seed (R5) at the 40% model threshold, and no application was made for the 50% model threshold. All fungicides were applied at 15 gal/acre and 40 psi using either a Lee self-propelled sprayer at R3 or a CO_2 backpack sprayer at R5 equipped with a 10-foot boom, fitted with six TJ-VS 8002 nozzles spaced 20 inches apart. Foliar disease ratings were rated on August 31. Septoria brown spot (SBS) and frogeye leaf spot (FLS) were rated for disease severity by visually assessing the percentage of symptomatic leaf area in the canopy. The two center rows of each plot were harvested on October 3, and yields were adjusted to 13% moisture. All data were analyzed in SAS 9.4 (SAS Institute, Cary, NC). A generalized linear mixed model analysis of variance was performed using PROC GLIMMIX. Values are least squares means, and values with different letters are significantly different based on Fisher's least significant difference ($\alpha = 0.05$).

In 2024, weather conditions were not favorable for disease development, and very little disease developed in plots. SBS and FLS were present in the trial, with SBS as the most prominent disease. All fungicide treatments significantly reduced SBS compared to nontreated control and 50% threshold with no application (Table 8). Application at both the R3 and at 30% threshold increased canopy greenness compared to the nontreated control but was not significantly different from application at 40% threshold or no application at 50%. No significant differences between treatments were observed for harvest moisture, test weight, and yield of soybean.

TABLE 8. *Effect of Treatments on Foliar Diseases, Canopy Greenness, and Yield of Soybean*

TREATMENT, RATE/ACRE, AND TIMING[z]	SBS SEVERITY[y] %	FLS SEVERITY[y] %	CANOPY GREEN[x] %	HARVEST MOISTURE %	TEST WEIGHT LB/BU	YIELD[w] BU/ACRE
Nontreated control	11.3 a	0.5	81.3 c	13.1	55.3	75.5
Revytek at 8.0 fl oz at R3	1.0 c	0.3	85.8 a	13.0	55.0	77.6
Revytek at 8.0 fl oz at Frogspotter 30% applied at R3	0.9 c	0.3	85.0 ab	12.9	54.9	77.8
Revytek at 8.0 fl oz at Frogspotter 40% applied at R5	1.5 c	0.1	82.5 bc	12.7	55.5	76.2
Revytek at 8.0 fl oz at Frogspotter 50% no application	6.3 b	0.4	83.8 abc	12.9	55.2	75.1
P-value[v]	0.0005	0.1366	0.0161	0.7529	0.7504	0.6836

[z] Fungicides were applied on July 18 for milk (R3) growth stages and 30% threshold and on August 1 at the dent (R5) growth stages for the 40% threshold, and no application was made for the 50% threshold. All treatments contained a nonionic surfactant (Preference) at a rate of 0.25% v/v.

[y] Foliar disease severity was rated on scale of 0–100% of canopy within a plot with disease symptoms on August 31. FLS was rated in the upper canopy, and SBS was rated in the lower canopy. SBS = Septoria brown spot. FLS = frogeye leaf spot.

[x] Canopy greenness was visually assessed on a scale of 0–100% green of the plot as a whole on August 31.

[w] Yields were adjusted to 13% moisture and harvested on October 3.

[v] All data were analyzed in SAS 9.4 (SAS Institute, Cary, NC). A generalized linear mixed model analysis of variance was performed using PROC GLIMMIX. Values are least squares means, and values with different letters are significantly different based on Fisher's least significant difference (α = 0.05).

EVALUATION OF SEED TREATMENT FOR PYTHIUM IN SOYBEAN IN CENTRAL INDIANA, 2024 (SOY24-13.ACRE)

E. R. Myers, S. Shim, and D. E. P. Telenko, Department of Botany and Plant Pathology, Purdue University, West Lafayette, IN 47907-2054

SOYBEAN (*GLYCINE MAX* LGS3253XF)

Seedling disease, *Pythium sylvaticum*

A trial was established at the Purdue Agronomy Center for Research and Education (ACRE) in Tippecanoe County, Indiana. The experiment was a randomized complete block design with four replications. Plots were 10 feet wide and 30 feet long and consisted of four rows, and the two center rows were used for evaluation. The previous crop was corn. Standard practices for soybean production in Indiana were followed. Soybean cultivar LGS3253XF was planted in 30-inch row spacing at a rate of 8 seeds/foot on May 13. Seed treatments were applied by cooperator. *Pythium sylvaticum* inoculum was applied at planting at 1.25 g/foot within the seedbed. Stand counts were assessed on July 1 at silk (R1) growth stage. The two center rows of each plot were harvested on October 3, and yields were adjusted to 13% moisture. All data were analyzed in SAS 9.4 (SAS Institute, Cary, NC). A generalized linear mixed model analysis of variance was performed using PROC GLIMMIX. Values are least squares means, and values with different letters are significantly different based on Fisher's least significant difference (α = 0.05).

In 2024, weather conditions were not favorable for disease development, and very little disease developed in plots. There was no significant effect of seed treatment on the stand count (Table 9). There was no significant difference between seed treatments on harvest moisture, test weight, or yield of soybean.

TABLE 9. *Effect of Treatment on Stand Count and Yield of Soybean*

TREATMENT[z]	STAND COUNT #/ACRE[y]	HARVEST MOISTURE %	TEST WEIGHT LB/BU	YIELD[x] BU/ACRE
Nontreated, inoculated control	99,970	12.8	54.2	71.3
Accerleron	118,483	12.3	54.5	72.7
Intego Suite	112,167	12.6	54.5	71.6
Zeltera Suite	107,158	13.2	54.0	74.7
Cruiser Maxx Vibrance	103,673	13.8	54.5	69.5
Cruiser Maxx APX (Vayantis)	107,593	12.9	54.1	72.2
P-value[w]	0.0845	0.1356	0.6040	0.4966

[z] Seed treatments were applied by the cooperator.
[y] Stand counts were assessed on July 1 at silk (R1) growth stage.
[x] Yields were adjusted to 13% moisture and harvested on October 3.
[w] All data were analyzed in SAS 9.4 (SAS Institute, Cary, NC). A generalized linear mixed model analysis of variance was performed using PROC GLIMMIX. Values are least squares means, and values with different letters are significantly different based on Fisher's least significant difference (α = 0.05).

PLANTING DATE AND SULFUR EVALUATION FOR SUDDEN DEATH SYNDROME ON SOYBEAN IN CENTRAL INDIANA, 2024 (SOY24-19.ACRE)

E. A. Duncan, S. Shim, S. Casteel, and D. E. P. Telenko, Department of Botany and Plant Pathology & Department of Agronomy, Purdue University, West Lafayette, IN 47907-2054

SOYBEAN (*GLYCINE MAX*)

Sudden death syndrome, *Fusarium virguliforme*

A trial was established at the Purdue Agronomy Center for Research and Education (ACRE) in Tippecanoe County, Indiana. The experimental design was a split plot with four replications. The main plot was planting date (April and May), and subplots were a factorial arrangement of inoculation (nontreated and inoculated) by treatment (nontreated control, ammonium sulfate, ammonium thiosulfate, and calcium sulfate). Plots were 10 feet wide and 30 feet long and consisted of four rows, and the two center rows were used for evaluation. The previous crop was corn. Standard practices for soybean production in Indiana were followed. Soybean seeds were planted in 30-inch row spacing at a rate of 140,00 seeds/acre. Soybeans were planted on April 18 (April planting) and on May 13 (May planting). *Fusarium virguliforme* was inoculated at planting at 1.25 g/foot. Sulfur treatments were applied on April 18 and May 14 following planting, with a resultant sulfur rate of 20 lb/acre. Ammonium sulfate 83 lb/acre and calcium sulfate at 117 lb/acre were hand-applied; ammonium thiosulfate was applied at 6.9 gal/acre at 15 gal/acre at 28–29 psi using a CO_2 backpack sprayer equipped with a 10-foot boom, fitted with eight TJ-VS 8002 nozzles spaced 15 inches apart at 3 mph. Disease ratings were assessed on August 30 at full seed (R6) growth stage. SDS in each plot was rated for disease incidence (DI) a percentage of plants with disease symptoms (0–100%), and disease severity (DS) was rated on a scale of 1 to 9 where 1 refers to low disease pressure and 9 refers to premature death of the plant. The SDS index (DX) was then calculated using the equation DX = (DI x DS/9). Canopy greenness was visually assessed on a scale of 0–100% green of the plot as a whole on August 30. The two center rows of each plot were harvested on October 3, and yields were adjusted to 13% moisture. All disease and yield data were analyzed in SAS 9.4 (SAS Institute, Cary, NC). A generalized linear mixed model analysis of variance was performed using PROC GLIMMIX. Values are least squares means, and values with different letters are significantly different based on Fisher's least significant difference (α = 0.05).

In 2024 weather conditions were not favorable for diseases, and very little disease developed in trial. SDS was present at a low severity. There was only one significant interaction between inoculum and treatment for yield; therefore, main effects are shown for simplicity (Table 10). There were no significant treatment effects for the SDS index. Canopy greenness at R6 was significantly higher for May 13 planted soybeans compared to April 18 planted soybeans, but there were no significant differences between inoculation or sulfur treatments. There were no significant effects of planting date or inoculation on harvest moisture, test weight, and soybean yield. Ammonium sulfate increased soybean yield over nontreated and ammonium thiosulfate but was not significantly different from calcium sulfate.

TABLE 10. *Effect of Planting Date, Inoculation, Sulfur Treatment on Sudden Death Syndrome and Yield of Soybean*

TREATMENTS AND RATE/ACRE[z]	SDS INDEX[y]	CANOPY GREEN[x] %	HARVEST MOISTURE %	TEST WEIGHT LB/BU	YIELD[w] BU/ACRE
April planting (April 18)	1.4	87.0 b	10.9	54.9	82.9
May planting (May 13)	0.6	93.9 a	10.9	54.9	82.0
Non-inoculated	1.1	90.2	10.9	55.0	82.7
Inoculated	0.8	90.7	11.0	54.8	82.2
Nontreated control	1.2	90.6	10.9	54.9	80.6 b
Ammonium sulfate 83.0 lb	0.7	90.5	11.0	55.0	85.6 a
Ammonium thiosulfate 6.9 gal	1.3	90.6	11.0	54.7	81.3 b
Calcium sulfate 117.0 lb	0.8	90.0	10.8	54.9	82.2 ab
P-value planting date[v]	*0.4677*	*0.0017*	*0.9124*	*0.9593*	*0.6005*
P-value inoculum	*0.2751*	*0.4102*	*0.3700*	*0.1895*	*0.6979*
P-value sulfur treatment	*0.2869*	*0.9002*	*0.2546*	*0.5942*	*0.0502*
P- value planting date*treatment	*0.9548*	*0.2635*	*0.7340*	*0.2606*	*0.9574*
P-value planting date*inoculum	*0.4248*	*0.7135*	*0.5066*	*0.8129*	*0.6153*
P-value inoculum*treatment	*0.3094*	*0.4396*	*0.6687*	*0.6429*	*0.0001*
P-value planting date*inoculum*treatment	*0.4470*	*0.8605*	*0.2919*	*0.8781*	*0.2364*

[z] *Fusarium virguliforme* grown on sorghum were inoculated at planting. Sulfur treatments were applied by hand following planting with a resultant sulfur rate of 20 lb/acre.o.

[y] SDS index was calculated using the equation DX = (DI x DS/9).

[x] Canopy greenness was visually assessed on a scale of 0–100% green of the plot as a whole on August 30.

[w] Yields were adjusted to 13% moisture and harvested on October 3.

[v] All data were analyzed in SAS 9.4 (SAS Institute, Cary, NC). A generalized linear mixed model analysis of variance was performed using PROC GLIMMIX. Values are least squares means, and values with different letters are significantly different based on Fisher's least significant difference (α = 0.05).

EVALUATION OF FUNGICIDES FOR FOLIAR DISEASES IN SOYBEAN IN CENTRAL INDIANA, 2024 (SOY24-22.ACRE)

S. Shim and D. E. P. Telenko, Department of Botany and Plant Pathology, Purdue University, West Lafayette, IN 47907-2054

SOYBEAN (*GLYCINE MAX* P29A19E)

Septoria brown spot, *Septoria glycines*
Frogeye leaf spot, *Cercospora sojina*

A trial was established at the Purdue Agronomy Center for Research and Education (ACRE) in Tippecanoe County, Indiana. The experiment was a randomized complete block design with four replications. Plots were 10 feet wide and 30 feet long and consisted of four rows, and the two center rows were used for evaluation. The previous crop was corn. Standard practices for soybean production in Indiana were followed. Soybean cultivar P29A19E was planted in 30-inch row spacing at a rate of 140,000 seeds/acre on May 8. All fungicides were applied at 15 gal/acre and 40 psi using a Lee self-propelled sprayer equipped with a 10-foot boom, fitted with six TJ-VS 8002 nozzles spaced 20 inches apart. Fungicides were applied on June 17 at V4 and on July 18 at the beginning pod (R3) growth stage. Foliar disease ratings were rated on August 30. Septoria brown spot (SBS) and frogeye leaf spot (FLS) were rated for disease severity by visually assessing the percentage of symptomatic leaf area in the canopy. The two center rows of each plot were harvested on September 26, and yields were adjusted to 13% moisture. All data were analyzed in SAS 9.4 (SAS Institute, Cary, NC). A generalized linear mixed model analysis of variance was performed using PROC GLIMMIX. Values are least squares means, and values with different letters are significantly different based on Fisher's least significant difference (α = 0.05).

In 2024, weather conditions were moderately favorable for disease development. SBS and FLS were present in the trial. All fungicide treatments significantly reduced SBS and FLS over nontreated control (Table 11). There was no significant effect of treatment on percentage of canopy greenness on August 30. There was no significant effect of treatment on harvest moisture, test weight, and yield of soybean.

TABLE 11. *Effect of Treatment on Foliar Disease Severity, Canopy Greenness, and Yield of Soybean*

TREATMENT, RATE/ACRE, AND TIMING[z]	SBS[y] %	FLS[y] %	CANOPY[x] GREEN %	HARVEST MOISTURE %	TEST WEIGHT LB/BU	YIELD[w] BU/ACRE
Nontreated control	10.3 a	1.1 a	82.5	15.0	55.2	65.3
Lucento 4.17 SC 5.0 fl oz at R3	0.9 b	0.2 b	85.0	15.1	55.2	65.6
Adastrio 4.0 SC 8.0 fl oz at R3	1.1 b	0.3 b	82.5	15.0	54.8	64.2
Topguard EQ 4.29 SC 7.0 fl oz at V4 fb Lucento 4.17 SC 5.0 fl oz at R3	1.3 b	0.2 b	82.5	15.2	54.6	66.9
Topguard EQ 4.29 SC 7.0 fl oz at V4 fb Adastrio 4.0 SC 8.0 fl oz at R3	1.3 b	0.3 b	86.3	14.7	53.8	63.4
Delaro Complete 3.82 SC 8.0 fl oz at R3	1.0 b	0.2 b	86.3	15.2	54.5	67.3
Revytek 4.44 SC 8.0 fl oz at R3	0.6 b	0.2 b	86.3	15.2	54.7	64.7
P-value[v]	0.0001	0.0080	0.5161	0.4763	0.3004	0.9546

[z] Fungicides were applied on June 17 at V4 and on July 18 at the beginning pod (R3) growth stage, and all treatments contained a nonionic surfactant (Preference) at a rate of 0.25% v/v. fb = followed by.

[y] Foliar disease severity was rated on a scale of 0–100% of canopy within a plot with disease symptoms. SBS = Septoria brown spot. FLS = frogeye leaf spot.

[x] Canopy greenness was visually assessed on a scale of 0–100% green within a plot.

[w] Yields were adjusted to 13% moisture and harvested on September 26.

[v] All data were analyzed in SAS 9.4 (SAS Institute, Cary, NC). A generalized linear mixed model analysis of variance was performed using PROC GLIMMIX. Values are least squares means, and values with different letters are significantly different based on Fisher's least significant difference (α = 0.05).

EVALUATION OF FUNGICIDES FOR FOLIAR DISEASES IN SOYBEAN IN CENTRAL INDIANA, 2024 (SOY24-27.ACRE)

S. Shim and D. E. P. Telenko, Department of Botany and Plant Pathology, Purdue University, West Lafayette, IN 47907-2054

SOYBEAN (*GLYCINE MAX* P29A19E)

Frogeye leaf spot, *Cercospora sojina*
Septoria brown spot, *Septoria glycines*

A trial was established at the Purdue Agronomy Center for Research and Education (ACRE) in Tippecanoe County, Indiana. The experiment was a randomized complete block design with four replications. Plots were 10 feet wide and 30 feet long and consisted of four rows, and the two center rows were used for evaluation. The previous crop was corn. Standard practices for soybean production in Indiana were followed. Soybean cultivar P29A19E was planted in 30-inch row spacing at a rate of 8 seeds/foot on May 13. Xyway LFR treatments were applied 2x2 at planting at 10 gal/acre. All foliar fungicides were applied at 15 gal/acre and 40 psi using a Lee self-propelled sprayer equipped with a 10-foot boom, fitted with six TJ-VS 8002 nozzles spaced 20 inches apart. Foliar fungicides were applied on July 18 at beginning pod (R3) growth stage. Foliar disease ratings were rated on August 30. Frogeye leaf spot (FLS) and Septoria brown spot (SBS) were rated for disease severity by visually assessing the percentage of symptomatic leaf area in the canopy. Percent canopy greenness was visually assessed as a percentage (0–100%) of plants in plot on September 26. The two center rows of each plot were harvested on September 26, and yields were adjusted to 13% moisture. All data were analyzed in SAS 9.4 (SAS Institute, Cary, NC). A generalized linear mixed model analysis of variance was performed using PROC GLIMMIX. Values are least squares means, and values with different letters are significantly different based on Fisher's least significant difference (α = 0.05).

In 2024, weather conditions were not favorable for disease development. SBS and FLS were present in the trial but only reached low levels. All treatments significantly reduced SBS and FLS over nontreated control (Table 12). There was no significant effect of treatment on percent canopy greenness on August 30. There was no significant effect of treatment on harvest moisture and yield of soybean. Xyway LFR 10.5 fl oz 2x2 had significantly higher test weight than Xyway LFR 10.5 fl oz 2x2 followed by Lucento 5.0 fl oz at R3 and Xyway LFR 10.5 fl oz 2x2 followed by Lucento 5.0 fl oz at R3 but was not significantly different from nontreated control or Xyway LFR 15.2 2x2. There was no significant effect of treatment on yield of soybean.

TABLE 12. *Effect of Treatment on Foliar Diseases, Canopy Greenness, and Yield of Soybean*

TREATMENT, RATE/ACRE, AND TIMING[z]	SBS[y] %	FLS[y] %	CANOPY GREEN[x] %	HARVEST MOISTURE %	TEST WEIGHT LB/BU	YIELD[w] BU/ACRE
Nontreated control	11.3 a	0.5 a	82.5	15.3	54.5 ab	67.1
Xyway LFR 10.5 fl oz 2x2 at plant	2.8 b	0.1 b	82.5	14.7	55.0 a	61.7
Xyway LFR 10.5 fl oz 2x2 at plant fb Lucento 5.0 fl oz at R3	0.7 b	0.0 b	82.5	14.8	54.4 bc	61.0
Xyway LFR 15.2 fl oz 2x2 at plant	3.3 b	0.1 b	82.5	15.5	54.8 ab	64.7
Xyway LFR 15.2 fl oz 2x2 at plant fb Lucento 5.0 fl oz at R3	0.4 b	0.0 b	85.0	15.1	54.0 c	61.6
P-value[v]	0.0002	0.0048	0.5165	0.5990	0.0155	0.5652

[z] Fungicides were applied in 2x2 at planting on May 13 and July 18 at the R3 (beginning pod) growth stage. fb = followed by.

[y] Foliar disease severity was rated on a scale of 0–100% of canopy within a plot with disease symptoms on August 30. SBS = Septoria brown spot; FLS = frogeye leaf spot.

[x] Canopy greenness was visually assessed on a scale of 0–100% green within a plot on August 30.

[w] Yields were adjusted to 13% moisture and harvested on September 26.

[v] All data were analyzed in SAS 9.4 (SAS Institute, Cary, NC). A generalized linear mixed model analysis of variance was performed using PROC GLIMMIX. Values are least squares means, and values with different letters are significantly different based on Fisher's least significant difference ($\alpha = 0.05$).

EVALUATION OF FOLIAR FUNGICIDE FOR FUSARIUM HEAD BLIGHT MANAGEMENT IN CENTRAL INDIANA, 2024 (WHT24-01.ACRE)

S. Shim and D.E.P. Telenko, Department of Botany and Plant Pathology, Purdue University, West Lafayette, IN 47907-2054

WHEAT (*TRITICUM AESTIVUM* P25R40)

Fusarium head blight, *Fusarium graminearum*

A trial was established at the Purdue Agronomy Center for Research and Education (ACRE) in Tippecanoe County, Indiana. The experiment was a randomized complete block design with four replications. Plots were 7.5 feet wide and 20 feet long and consisted of 12 rows spaced 7.5 inches apart, and the center of each plot was used for evaluation. The previous crop was corn. On October 26, 2023, wheat cultivar P25R40 was drilled at 7.5-inch spacing. All fungicide applications were applied at 15 gal/acre and 40 psi using a CO_2 backpack sprayer equipped with a 10-foot boom, fitted with six TJ-VS 8002 nozzles spaced 20 inches apart and directed forward and backward at a 45-degree angle. Fungicides were applied on May 15 and May 20 at Feekes growth stages 10.5.1 and 10.5.1 + 5 days, respectively. All plots were inoculated with a mixture of isolates of *Fusarium graminearum* endemic to Indiana on May 15. The spore suspension (50,000 spores/ml) was applied at 300 ml/plot with the CO_2 backpack sprayer. Disease ratings were assessed on May 29. Fusarium head blight (FHB) incidence was measured as the number of infected heads out of 60 plants in each plot and calculated as a percentage. FHB severity was rated by visually assessing the percentage of the infected head, and the FHB index was calculated as % FHB incidence multiplied by average FHB severity/100 per plot. Values for each plot were averaged before analysis. The eight center rows of each plot were harvested with a Kincaid plot combine on June 28, and yields were adjusted to 13.5% moisture. A subsample of grain was taken from each plot and partitioned for deoxynivalenol (DON) analysis completed by the University of Minnesota DON testing lab and to determine Fusarium damaged kernels (FDK) by visually assessing the percentage (0–100%) of the infected heads. All data were analyzed in SAS 9.4 (SAS Institute, Cary, NC). A generalized linear mixed model analysis of variance was performed using PROC GLIMMIX. Values are least squares means, and values with different letters are significantly different based on Fisher's least significant difference ($\alpha = 0.05$).

In 2024, weather conditions were moderately favorable for FHB. FHB incidence, severity, and index were reduced by all fungicide treatments over the nontreated control except Miravis Ace 5.2SC at 10.5.1 for FHB incidence and Prosaro 421SC at 10.5.1 for FHB severity, but these were not significantly different from all other fungicide treatments or timings (Table 13). The percent of FDK visual was significantly reduced by all fungicide programs over nontreated control except Prosaro 421SC, Miravis Ace 5.2SC, and Miravis Ace 5.2SC followed by Tebuconazole 10.5.1 +5 days. The concentration of DON was significantly reduced over nontreated control by all treatments. There was no significant difference between treatments for yield of wheat.

TABLE 13. *Effect of Fungicide on Fusarium Head Blight (FHB), Fusarium Damaged Kernels (FDK), Deoxynivalenol (DON), and Yield of Wheat*

TREATMENT, RATE/ACRE, AND TIMING[z]	FHB INCIDENCE %[y]	FHB SEVERITY %[x]	FHB INDEX[w]	FDK[v] %	DON[u] PPM	YIELD[t] BU/ACRE
Nontreated control	25.8 a	13.0 a	3.5 a	11.8 a	4.0 a	77.8
Prosaro 421SC 6.5 fl oz at 10.5.1	12.1 bc	9.5 ab	1.4 b	9.5 ab	1.8 bc	82.9
Miravis Era 13.5 fl oz at 10.5.1	12.1 bc	7.5 bc	1.1 b	5.8 c	0.8 c	84.1
Miravis Ace 5.2SC 13.7 fl oz at 10.5.1	17.1 ab	7.2 bc	1.4 b	9.8 ab	2.4 b	86.0
Prosaro Pro 400SC 10.3 fl oz at 10.5.1	7.5 bc	6.3 bc	0.7 b	8.5 bc	1.1 c	89.8
Sphaerex 2.50SC 7.3 fl oz at 10.5.1	10.8 bc	7.8 bc	0.8 b	7.0 bc	0.9 c	85.5
Miravis Ace 5.2SC 13.7 fl oz at 10.5.1 fb Prosaro Pro 400SC 10.3 fl oz at 10.5.1 +5d	7.1 c	4.0 c	0.3 b	7.5 bc	0.8 c	84.9
Miravis Ace 5.2SC 13.7 fl oz at 10.5.1 fb Sphaerex 2.50SC 7.3 fl oz at 10.5.1 +5d	7.9 bc	4.7 bc	0.4 b	8.0 bc	0.9 c	83.4
Miravis Ace 5.2SC 13.7 fl oz at 10.5.1 fb Tebuconazole 4.0 fl oz at 10.5.1 +5d	11.7 bc	4.6 c	0.6 b	9.3 ab	1.3 c	94.7
P-value[s]	0.0155	0.0203	0.0025	0.0268	0.0001	0.3912

[z] Fungicide treatments applied on May 15 and May 20 at Feekes growth stage 10.5.1 and 10.5.1 + 5 days, respectively. All treatments contained a nonionic surfactant (Preference) at a rate of 0.125% v/v. All plots were inoculated with a mixture of isolates of *Fusarium graminearum* endemic to Indiana on May 23, with a spore suspension (50,000 spores/ml) applied at 300 ml/plot with handheld sprayer on May 15. fb = followed by.

[y] FHB incidence was measured as the number of infected heads out of 60 plants in each plot and calculated as a percentage on May 29.

[x] FHB severity was rated by visually assessing the percentage of the infected head.

[w] FHB index was calculated as % FHB incidence multiplied by average FHB severity/100 per plot.

[v] Visual assessment of percentage of Fusarium damaged kernels (FDK) was performed.

[u] Analysis of the mycotoxin DON was completed by the University of Minnesota DON Testing Lab.

[t] Yields were adjusted to 13.5% moisture and harvested on June 28.

[s] All data were analyzed in SAS 9.4 (SAS Institute, Cary, NC). A generalized linear mixed model analysis of variance was performed using PROC GLIMMIX. Values are least squares means, and values with different letters are significantly different based on Fisher's least significant difference (α = 0.05).

INTEGRATED MANAGEMENT OF FUSARIUM HEAD BLIGHT OF WHEAT IN CENTRAL INDIANA, 2024 (WHT24-02.ACRE)

C. Rocco da Silva, S. Shim and D.E.P. Telenko, Department of Botany and Plant Pathology, Purdue University, West Lafayette, IN 47907-2054

WHEAT (*TRITICUM AESTIVUM* P25R40 AND P25R61)

Fusarium head blight, *Fusarium graminearum*

A trial was established at the Purdue Agronomy Center for Research and Education (ACRE) in Tippecanoe County, Indiana. The experiment was a randomized complete block design with four replications. Plots were 7.5 feet wide and 20 feet long and consisted of 12 rows spaced 7.5 inches apart, and the center of each plot was used for evaluation. The previous crop was corn. On October 26, 2023, wheat cultivars P25R40 and P25R61 were drilled at 7.5-inch spacing. All fungicide applications were applied at 15 gal/acre and 40 psi using a CO_2 backpack sprayer equipped with a 10-foot boom, fitted with six TJ-VS 8002 nozzles spaced 20 inches apart and directed forward and backward at a 45-degree angle. Fungicides were applied on May 15 at Feekes growth stage 10.5.1. All plots were inoculated with a mixture of isolates of *Fusarium graminearum* endemic to Indiana on May 16. The spore suspension (50,000 spores/ml) was applied at 300 ml/plot with the CO_2 backpack sprayer. Disease ratings were assessed on May 29. Fusarium head blight (FHB) incidence was measured as the number of infected heads out of 60 plants in each plot and calculated as a percentage. FHB severity was rated by visually assessing the percentage of the infected head, and the FHB index was calculated as % FHB incidence multiplied by average FHB severity/100 per plot. Values for each plot were averaged before analysis. The eight center rows of each plot were harvested with a Kincaid plot combine on June 28, and yields were adjusted to 13.5% moisture. A subsample of grain was taken from each plot and partitioned for deoxynivalenol (DON) analysis completed by the University of Minnesota DON testing lab and to determine Fusarium damaged kernels (FDK) by visually assessing the percentage (0–100%) of the infected heads. All data were analyzed in SAS 9.4 (SAS Institute, Cary, NC). A generalized linear mixed model analysis of variance was performed using PROC GLIMMIX. Values are least squares means, and values with different letters are significantly different based on Fisher's least significant difference ($\alpha = 0.05$).

In 2024, weather conditions were moderately favorable for FHB. FHB was the most prominent disease in the trial. The main effects of cultivar and fungicide treatment are presented (Table 14). FHB index, DON, and yield were lowest in the resistant cultivar P25R61. Test weight was significantly higher in P25R40 as compared to P25R61. The FHB index was reduced by all fungicide treatments over nontreated, non-inoculated control, but Prosaro was not significant from nontreated, inoculated control. Applications of Prosaro 421SC, Prosaro Pro 400SC, and Sphaerex 2.50SC resulted in the lowest percent FDK. The concentration of DON was significantly reduced by all the fungicides over the nontreated, inoculated, and non-inoculated controls. All fungicide treatments increased harvest moisture and test weight over the nontreated controls except Prosaro and nontreated, inoculated control. There were no significant differences in treatment for yield of wheat.

TABLE 14. *Effect of Cultivar and Fungicide on Fusarium Head Blight (FHB), Fusarium Damaged Kernels (FDK), Deoxynivalenol (DON), and Yield of Wheat*

TREATMENT AND RATE/ACRE[z]	FHB[y] INDEX	FDK[x] %	DON[w] PPM	MOISTURE %	TEST WEIGHT LB/BU	YIELD[v] BU/ACRE
Cultivar						
P25R40 (susceptible)	1.6 a	11.4	2.7 a	16.4	56.1 a	86.6 a
P25R61 (resistant)	0.8 b	12.2	1.1 b	16.4	55.3 b	66.8 b
Fungicide						
Nontreated, inoculated control	1.5 ab	14.1 ab	3.6 a	16.2 b	55.4 bc	70.9
Nontreated, non-inoculated control	2.1 a	15.8 a	3.8 a	16.1 b	55.1 c	79.4
Prosaro 421SC 6.5 fl oz	1.1 bc	9.5 cd	1.2 b	16.5 a	55.8 ab	77.8
Miravis Ace 5.2SC 13.7 fl oz	0.9 c	12.5 bc	1.5 b	16.5 a	55.9 a	76.1
Prosaro Pro 400SC 10.3 fl oz	0.8 c	10.4 cd	1.2 b	16.6 a	56.1 a	75.1
Sphaerex 2.50SC 7.3 fl oz	0.8 c	8.5 d	1.0 b	16.6 a	56.0 a	81.0
P-value cultivar[u]	*0.0001*	*0.4081*	*0.0001*	*0.3427*	*0.0001*	*0.0001*
P-value treatment	*0.0005*	*0.0002*	*0.0001*	*0.0001*	*0.0014*	*0.6317*
P-value cultivar*treatment	*0.4272*	*0.1842*	*0.0128*	*0.8080*	*0.1615*	*0.7108*

[z] Fungicide treatments applied on May 15 at Feekes growth stage 10.5. 1 and contained a nonionic surfactant (Preference) at a rate of 0.125% v/v. All plots were inoculated with *Fusarium graminearum* spore suspension (50,000 spores/ml) after the treatment at Feekes 10.5.1. Spore suspension was applied at 300 ml/plot with handheld sprayer on May 16.

[y] The FHB index was calculated as % FHB incidence multiplied by average FHB severity/100 per plot.

[x] Visual assessment of percentage (0–100%) of Fusarium damaged kernels (FDK) was performed.

[w] Analysis of the mycotoxin DON was completed by the University of Minnesota DON Testing Lab.

[v] Yields were adjusted to 13.5% moisture and harvested on June 28.

[u] All data were analyzed in SAS 9.4 (SAS Institute, Cary, NC). A generalized linear mixed model analysis of variance was performed using PROC GLIMMIX. Values are least squares means, and values with different letters are significantly different based on Fisher's least significant difference (α = 0.05).

EVALUATION OF FOLIAR FUNGICIDES FOR WHEAT DISEASE MANAGEMENT IN CENTRAL INDIANA, 2024 (WHT24-05.ACRE)

S. Shim and D.E.P. Telenko, Department of Botany and Plant Pathology, Purdue University, West Lafayette, IN 47907-2054

WHEAT (*TRITICUM AESTIVUM* P25R40)

Fusarium head blight, *Fusarium graminearum*
Leaf blotch, *Septoria tritici/Stagnospora nodorum*
Leaf rust, *Puccinia triticina*

A trial was established at the Purdue Agronomy Center for Research and Education (ACRE) in Tippecanoe County, Indiana. The experiment was a randomized complete block design with four replications. Plots were 7.5 feet wide and 20 feet long and consisted of 12 rows spaced 7.5 inches apart, and the center of each plot was used for evaluation. The previous crop was corn. On October 26, 2023, wheat cultivar P25R40 was drilled at 7.5-inch spacing. All fungicide applications were applied at 15 gal/acre and 40 psi using a CO_2 backpack sprayer equipped with a 10-foot boom, fitted with six TJ-VS 8002 nozzles spaced 20 inches apart and directed forward and backward at a 45-degree angle. Fungicides were applied on April 30 at Feekes growth stage 8 and May 15 at Feekes growth stage 10.5.1. Disease ratings were assessed on May 29. Fusarium head blight (FHB) incidence was measured as the number of infected heads out of 60 plants in each plot and calculated as a percentage. FHB severity was rated by visually assessing the percentage of the infected head, and the FHB index was calculated as % FHB incidence multiplied by average FHB severity/100 per plot. Values for each plot were averaged before analysis. The eight center rows of each plot were harvested with a Kincaid plot combine on June 28, and yields were adjusted to 13.5% moisture. A subsample of grain was taken from each plot and partitioned for deoxynivalenol (DON) analysis completed by the University of Minnesota DON testing lab and to determine Fusarium damaged kernels (FDK) by visually assessing the percentage (0–100%) of the infected heads. All data were analyzed in SAS 9.4 (SAS Institute, Cary, NC). A generalized linear mixed model analysis of variance was performed using PROC GLIMMIX. Values are least squares means, and values with different letters are significantly different based on Fisher's least significant difference (α = 0.05).

In 2024, weather conditions were moderately favorable for FHB, but very little foliar disease developed in trial. There were no significant differences between treatments for leaf blotch (Stag/Sept) and the FHB index (Table 15). All treatments reduced leaf rust severity compared to the nontreated control. No significant differences were detected between treatments for FDK and wheat yield. The concentration of DON was reduced only by the application of Adastrio 3.5SC at Feekes 8 growth stage.

TABLE 15. *Effect of Fungicide on Fusarium Head Blight (FHB), Fusarium Damaged Kernels (FDK), Deoxynivalenol (DON), and Yield of Wheat*

TREATMENT, RATE/ACRE, AND TIMING[z]	STAG/SEPT % SEVERITY[y]	RUST % SEVERITY[y]	FHB INDEX[x]	FDK %[w]	DON PPM[v]	YIELD[u] BU/ACRE
Nontreated control	0.1	0.1 a	2.2	1.5	5.9 a	59.6
Nexicor 2.96 EC 7.0 fl oz at Feekes 8	0.1	0.0 b	1.4	4.3	5.8 a	62.5
Topguard 1.04 SC 10.0 fl oz at Feekes 8	0.0	0.0 b	2.0	2.0	5.3 a	63.2
Priaxor 4.17 EC 4.0 fl oz at Feekes 8	0.1	0.0 b	1.3	2.0	4.7 a	62.3
Trivapro 2.21 SE 9.4 fl oz at Feekes 8	0.1	0.0 b	1.1	2.5	5.0 a	57.9
Delaro 325 SC 8.0 fl oz at Feekes 8	0.0	0.0 b	2.2	1.0	4.5 a	59.6
Quilt Xcel 2.2 SE 10.5 fl oz at Feekes 8	0.1	0.0 b	1.5	2.5	5.9 a	58.9
Tilt 3.6 ES 4.0 fl oz at Feekes 8	0.1	0.0 b	2.3	6.0	5.4 a	56.4
Headline SC 6.0 fl oz at Feekes 8	0.1	0.0 b	2.0	2.0	5.3 a	57.8
Adastrio 3.5 SC 6.0 fl oz at Feekes 8	0.0	0.0 b	0.8	0.9	2.1 b	66.7
Prosaro 1.76 SC 6.5 fl oz at Feekes 10.5.1	0.2	0.0 b	1.9	2.0	5.8 a	56.7
P-value[t]	*0.9319*	*0.0096*	*0.7540*	*0.2035*	*0.0091*	*0.1226*

[z] Fungicides were applied on April 30 at Feekes growth stage 8 and May 15 at Feekes growth stage 10.5.1. All treatments contained a nonionic surfactant (Preference) at a rate of 0.125% v/v.

[y] Foliar disease severity was rated by visually assessing the percentage (0–100%) of symptomatic leaf area on the flag leaf of five leaves per plot. Stag/Sept = *Stagnospora nodorum/Septoria tritici* (leaf blotch).

[x] The FHB index was calculated as % FHB incidence multiplied by average FHB severity/100 per plot.

[w] Visual assessment of percentage (0–100%) of Fusarium damaged kernels (FDK) was performed on June 12.

[v] Analysis of the mycotoxin DON was completed by the University of Minnesota DON Testing Lab.

[u] Yields were adjusted to 13.5% moisture and harvested on June 28.

[t] All data were analyzed in SAS 9.4 (SAS Institute, Cary, NC). A generalized linear mixed model analysis of variance was performed using PROC GLIMMIX. Values are least squares means, and values with different letters are significantly different based on Fisher's least significant difference ($\alpha = 0.05$).

EVALUATION OF BIOLOGICAL FUNGICIDES FOR FUSARIUM HEAD BLIGHT OF WHEAT IN CENTRAL INDIANA, 2024 (WHT24-07.ACRE)

S. Shim and D. E. P. Telenko, Department of Botany and Plant Pathology, Purdue University, West Lafayette, IN 47907-2054

WHEAT (*TRITICUM AESTIVUM* P25R40)

Fusarium head blight, *Fusarium graminearum*

A trial was established at the Purdue Agronomy Center for Research and Education (ACRE) in Tippecanoe County, Indiana. The experiment was a randomized complete block design with four replications. Plots were 7.5 feet wide and 20 feet long and consisted of 12 rows spaced 7.5 inches apart, and the center of each plot was used for evaluation. The previous crop was corn. On October 26, 2023, wheat cultivar P25R40 was drilled at 7.5-inch spacing. All fungicide applications were applied at 15 gal/acre and 40 psi using a CO_2 backpack sprayer equipped with a 10-foot boom, fitted with six TJ-VS 8002 nozzles spaced 20 inches apart and directed forward and backward at a 45-degree angle. Fungicides were applied on May 15 at Feekes growth stage 10.5.1. All plots were inoculated with a mixture of isolates of *Fusarium graminearum* endemic to Indiana on May 15. The spore suspension (50,000 spores/ml) was applied at 300 ml/plot with the CO_2 backpack sprayer. Disease ratings were assessed on May 29. Fusarium head blight (FHB) incidence was measured as the number of infected heads out of 60 plants in each plot and calculated as a percentage. FHB severity was rated by visually assessing the percentage of the infected head. The HB index was calculated as % FHB incidence multiplied by average FHB severity/100 per plot. Values for each plot were averaged before analysis. The eight center rows of each plot were harvested with a Kincaid plot combine on June 28, and yields were adjusted to 13.5% moisture. A subsample of grain was taken from each plot and partitioned for deoxynivalenol (DON) analysis completed by the University of Minnesota DON testing lab and to determine Fusarium damaged kernels (FDK) by visually assessing the percentage (0–100%) of the infected heads. All data were analyzed in SAS 9.4 (SAS Institute, Cary, NC). A generalized linear mixed model analysis of variance was performed using PROC GLIMMIX. Values are least squares means, and values with different letters are significantly different based on Fisher's least significant difference (α = 0.05).

In 2024, weather conditions were moderately favorable for FHB. FHB incidence was significantly increased with Pacesetter over nontreated control (Table 16). No significant differences between treatments were detected for FHB severity. The FHB index was significantly higher with Pacesetter over other treatments but not significantly different from the nontreated control. The concentration of DON was significantly reduced by Prosaro application as compared to the nontreated control. There were no differences detected between treatments for FDK and wheat yield.

TABLE 16. *Effect of Fungicide on Fusarium Head Blight (FHB), Fusarium Damaged Kernels (FDK), Deoxynivalenol (DON), and Wheat Yield*

TREATMENT AND RATE/ACRE[z]	FHB % INCIDENCE[y]	FHB % SEVERITY[x]	FHB INDEX[w]	FDK %[v]	DON (PPM)[u]	YIELD[t] BU/ACRE
Nontreated control	14.6 bc	8.4	1.3 ab	1.5	4.0 a	64.4
Prosaro 421SC 8.2 fl oz	8.8 c	6.5	0.6 b	0.6	1.3 b	71.9
ChampION 50 WP 1.5 lb	15.9 ab	7.2	1.2 b	2.5	3.8 a	72.9
Pacesetter WS 13.0 fl oz	22.5 a	8.7	2.0 a	2.0	4.3 a	71.8
Sonata 1.0 qt	13.3 bc	8.6	1.1 b	1.9	3.7 a	66.8
Actinovate AG 12.0 fl oz	17.1 ab	6.5	1.1 b	2.0	4.1 a	81.0
P-value[s]	0.0204	0.4073	0.0472	0.2309	0.0107	0.0630

[z] Fungicide treatments applied on May 15 at Feekes growth stage 10.5.1. All plots were inoculated with a mixture of isolates of *Fusarium graminearum* endemic to Indiana, with a spore suspension (50,000 spores/ml) applied at 300 ml/plot with a handheld sprayer on May 15.

[y] FHB incidence was measured as the number of infected heads out of 60 plants in each plot and calculated as a percentage on May 29.

[x] FHB severity was rated by visually assessing the percentage of the infected head.

[w] The FHB index was calculated as % FHB incidence multiplied by average FHB severity/100 per plot.

[v] Visual assessment of percentage of Fusarium damaged kernels (FDK) was performed on July 11.

[u] Analysis of the mycotoxin DON was completed by the University of Minnesota DON Testing Lab.

[t] Yields were adjusted to 13.5% moisture and harvested on June 28.

[s] All data were analyzed in SAS 9.4 (SAS Institute, Cary, NC). A generalized linear mixed model analysis of variance was performed using PROC GLIMMIX. Values are least squares means, and values with different letters are significantly different based on Fisher's least significant difference ($\alpha = 0.05$).

EFFICACY OF ADASTRIO AND TOPGUARD FOR FOLIAR WHEAT DISEASE IN CENTRAL INDIANA, 2024 (WHT24-08.ACRE)

S. Shim and D. E. P. Telenko, Department of Botany and Plant Pathology, Purdue University, West Lafayette, IN 47907-2054

WHEAT (*TRITICUM AESTIVUM* P25R40)

Leaf blotch, *Septoria tritici/Stagnospora nodorum*

A trial was established at the Purdue Agronomy Center for Research and Education (ACRE) in Tippecanoe County, Indiana. The experiment was a randomized complete block design with four replications. Plots were 7.5 feet wide and 20 feet long and consisted of 12 rows spaced 7.5 inches apart, and the center of each plot was used for evaluation. The previous crop was corn. On October 26, 2023, wheat cultivar P25R40 was drilled at 7.5-inch spacing. All fungicide applications were applied at 15 gal/acre and 40 psi using a CO_2 backpack sprayer equipped with a 10-foot boom, fitted with six TJ-VS 8002 nozzles spaced 20 inches apart and directed forward and backward at a 45-degree angle. Fungicides were applied on March 21 and May 15 at greenup and Feekes 10.5.1, respectively. Disease ratings were assessed on May 29. Disease severity on leaves were rated by visually assessing the percentage of symptomatic leaf tissue on five flag leaves per plot for leaf blotch. Values for each plot were averaged before analysis. The eight center rows of each plot were harvested with a Kincaid plot combine on June 28, and yields were adjusted to 13.5% moisture. A subsample of grain was taken from each plot and partitioned for deoxynivalenol (DON) analysis completed by the University of Minnesota DON testing lab and to determine FDK by visually assessing the percentage (0–100%) of the infected heads. All data were analyzed in SAS 9.4 (SAS Institute, Cary, NC). A generalized linear mixed model analysis of variance was performed using PROC GLIMMIX. Values are least squares means, and values with different letters are significantly different based on Fisher's least significant difference (α = 0.05).

In 2024, weather conditions were not favorable for foliar diseases. Low levels of leaf blotch were detected. There were no differences in treatments from nontreated control for severity of leaf blotch (Stag/Sept) and FDK (Table 17). The concentration of DON was significantly reduced over nontreated control by all treatments except Adastrio applied at greenup. All treatments significantly increased harvest moisture over nontreated control except Adastrio applied at greenup. There were no significant differences between treatments for test weight and yield of wheat.

TABLE 17. *Effect of Fungicide on Foliar Diseases, Fusarium Damaged Kernels (FDK), Deoxynivalenol (DON), and Wheat Yield*

TREATMENT, RATE/ACRE, AND TIMING[z]	STAG/SEPT[y] % SEVERITY	FDK[x] %	DON[w] PPM	MOISTURE %	TEST WEIGHT LB/BU	YIELD[v] BU/ACRE
Nontreated control	0.1	1.5	6.0 a	15.1 b	55.5	58.0
Adastrio 4.0SC 3.5 fl oz at greenup	0.3	2.0	6.5 a	15.1 b	56.7	53.0
Adastrio 4.0SC 3.5 fl oz at greenup fb Adastrio 4.0SC 3.5 fl oz at Feekes 10.5.1	0.1	0.9	3.4 bc	15.6 a	56.6	61.3
Topguard SC 4.29 SC 5.0 fl oz at greenup fb Adastrio 4.0SC 5.0 fl oz at Feekes 10.5.1	0.0	0.8	2.5 c	15.8 a	56.5	60.4
Topguard SC 4.29 SC 5.0 fl oz at greenup fb Topguard SC 4.29 SC 5.0 fl oz at Feekes 10.5.1	0.2	1.4	3.7 b	15.6 a	56.0	55.4
P-value[u]	*0.2487*	*0.3441*	*0.0001*	*0.0036*	*0.5580*	*0.2073*

[z] Fungicide treatments applied on March 21 at greenup and May 15 at Feekes 10.5.1. All treatments contained a nonionic surfactant (Preference) at a rate of 0.25% v/v. fb = followed by.

[y] Foliar disease severity was rated by visually assessing the percentage of symptomatic leaf tissues on five flag leaves per plot on May 29. Stag/Sept = *Stagnospora nodorum/Septoria tritici* (leaf blotch).

[x] Visual assessment of percentage of Fusarium damaged kernels (FDK) was performed on a subsample (0–100%).

[w] Analysis of the mycotoxin DON was completed by the University of Minnesota DON Testing Lab.

[v] Yields were adjusted to 13.5% moisture and harvested on June 28.

[u] All data were analyzed in SAS 9.4 (SAS Institute, Cary, NC). A generalized linear mixed model analysis of variance was performed using PROC GLIMMIX. Values are least squares means, and values with different letters are significantly different based on Fisher's least significant difference (α = 0.05).

PINNEY PURDUE AGRICULTURAL CENTER (PPAC)

UNIFORM FUNGICIDE COMPARISON FOR TAR SPOT IN CORN IN NORTHWESTERN INDIANA, 2024 (COR24-02.PPAC)

M. Mizuno, S. Shim, and D. E. P. Telenko, Department of Botany and Plant Pathology, Purdue University, West Lafayette, IN 47907-2054

CORN (*ZEA MAYS* W2584 VT2P RIB)

Tar spot, *Phyllachora maydis*

A trial was established at the Pinney Purdue Agricultural Center (PPAC) in Porter County, Indiana. The experiment was a randomized complete block design with four replications. Plots were 10 feet wide and 30 feet long and consisted of four rows, and the two center rows were used for evaluation. The previous crop was corn. Standard practices for grain corn production in Indiana were followed. Corn hybrid W2584VT2P RIB was planted in 30-inch row spacing at a rate of 34,000 seeds/acre on May 23. The field was overhead irrigated weekly at 1 inch unless weekly rainfall was 1 inch or higher to encourage disease. All foliar fungicide applications were applied at 15 gal/acre and 40 psi using a Lee self-propelled sprayer equipped with a 10-foot boom, fitted with six TJ-VS 8002 nozzles spaced 20 inches apart. Fungicides were applied on August 7 at silking (R1) growth stage and three weeks after treatments (WAT) on August 29 at milk (R3) growth stage. Disease ratings were assessed on September 16 at dent (R5) growth stage. Tar spot was rated by visually assessing the percentage of stromata per leaf (0–100%) on five plants in each plot at the ear leaf (EL) and ear leaf ±2 (EL±2). The two center rows of each plot were harvested on October 21, and yields were adjusted to 15.5% moisture. All disease and yield data were analyzed in SAS 9.4 (SAS Institute, Cary, NC). A generalized linear mixed model analysis of variance was performed using PROC GLIMMIX. Values are least squares means, and values with different letters are significantly different based on Fisher's least significant difference (α = 0.05).

In 2024, weather conditions were favorable for disease. Tar spot was the most prominent disease in the trial and reached high severity. All treatments significantly reduced tar spot stromata severity compared to the non-treated control on September 16 at the EL and EL±2 (Table 18). On September 16, Aproach Prima followed by Headline AMP and Aproach Prima resulted in the lowest level of disease at the EL–2 but were only significantly different from Headline AMP followed by Veltyma and Headline AMP followed by Delaro Complete programs. At the EL, Delaro Complete followed by Headline AMP resulted in the lowest level of tar spot but was not significantly different from all other treatments except the single applications of Veltyma, Aproach Prima, Miravis Neo, and Headline AMP followed by Headline Amp. At the EL+2 all treatments led to 3% or less of tar spot compared to 20% in the nontreated control. There was no significant effect of treatment on harvest moisture and yield of corn.

TABLE 18. *Effect of Fungicide Programs on Tar Spot Stromata Severity, Canopy Greenness, and Yield of Corn*

TREATMENT, RATE/ACRE, AND TIMING[z]	TAR SPOT[y] % EL-2	TAR SPOT[y] % EL	TAR SPOT[y] % EL+2	HARVEST MOISTURE %	YIELD[x] BU/ACRE
Nontreated control	23.5 a	23.0 a	20.0 a	16.1	182.3
Veltyma 3.34 S 7.0 fl oz at R1	10.1 b-e	4.4 b-e	2.6 bc	17.1	217.0
Aproach Prima 2.34 SC 6.8 fl oz at R1	5.5 e	5.8 bc	2.6 bc	16.6	209.3
Miravis Neo 2.5 SE 13.7 fl oz at R1	6.3 de	6.0 b	3.1 b	18.4	194.1
Delaro Complete 3.82 SC 8.0 fl oz at R1	7.6 cde	3.6 b-f	2.0 b-e	19.2	195.5
Headline AMP 1.68 SC 10.0 fl oz at R1	9.8 b-e	5.1 bcd	2.5 bcd	18.4	189.4
Veltyma 3.34 S 7.0 fl oz at R1 fb Headline AMP 1.68 SC 10.0 fl oz 3 WAT	9.4 b-e	3.7 b-f	0.9 cde	19.4	201.2
Aproach Prima 2.34 SC 6.8 fl oz at R1 fb Headline AMP 1.68 SC 10.0 fl oz 3 WAT	5.6 e	1.9 ef	0.6 de	17.9	205.5
Miravis Neo 2.5 SE 13.7 fl oz at R1 fb Headline AMP 1.68 SC 10.0 fl oz 3 WAT	8.8 cde	2.6 def	0.5 e	19.6	197.5
Delaro Complete 3.82 SC 8.0 fl oz at R1 fb Headline AMP 1.68 SC 10.0 fl oz 3 WAT	8.3 cde	1.5 f	0.6 e	20.1	193.6
Headline AMP 1.68 SC 10.0 fl oz at R1 fb Veltyma 3.34 S 7.0 fl oz 3 WAT	11.8 bc	3.3 c-f	1.3 b-e	18.8	198.3
Headline AMP 1.68 SC 10.0 fl oz at R1 fb Aproach Prima 2.34 SC 6.8 fl oz 3 WAT	9.5 b-e	3.9 b-f	1.1 cde	17.9	204.9
Headline AMP 1.68 SC 10.0 fl oz at R1 fb Miravis Neo 2.5 SE 13.7 fl oz 3 WAT	8.0 cde	2.1 ef	0.9 cde	19.0	197.8
Headline AMP 1.68 SC 10.0 fl oz at R1 fb Delaro Complete 3.82 SC 8.0 fl oz 3 WAT	13.2 b	2.9 def	1.2 cde	18.4	201.6
Headline AMP 1.68 SC 10.0 fl oz at R1 fb Headline AMP 1.68 SC 10.0 fl oz 3 WAT	10.5 bcd	4.3 b-e	1.8 b-e	18.4	202.7
P-value[w]	*0.0001*	*0.0001*	*0.0001*	*0.2371*	*0.1684*

[z] Fungicides were applied on August 7 at blister (R1) growth stage and on August 29, 3 weeks after treatment (3 WAT), at milk (R3) growth stage. All treatments applied contained a nonionic surfactant (Preference) at a rate of 0.25% v/v. fb = followed by.

[y] Tar spot stromata severity was visually assessed as a percentage (0–100%) of leaf area on five plants in each plot on September 16 at dent (R5) growth stage at the ear leaf (EL) and ear leaf ±2 (EL±2).

[x] Yields were adjusted to 15.5% moisture and harvested on October 21.

[w] All data were analyzed in SAS 9.4 (SAS Institute, Cary, NC). A generalized linear mixed model analysis of variance was performed using PROC GLIMMIX. Values are least squares means, and values with different letters are significantly different based on Fisher's least significant difference (α = 0.05).

EVALUATION OF HYBRID AND FUNGICIDE TIMING FOR TAR SPOT IN CORN IN NORTHWESTERN INDIANA, 2024 (COR24-03.PPAC)

K. M. Goodnight and D. E. P. Telenko, Department of Botany and Plant Pathology, Purdue University, West Lafayette, IN 47907-2054

CORN (*ZEA MAYS* W2585VT2P AND P0589AMXT)

Tar spot, *Phyllachora maydis*

A trial was established at the Pinney Purdue Agricultural Center (PPAC) in Porter County, Indiana. The experiment was a randomized complete block design with four replications. Plots were 10 feet wide and 30 feet long and consisted of four rows, and the two center rows were used for evaluation. The previous crop was corn. Standard practices for grain corn production in Indiana were followed. Corn hybrids W2585VT2P and P0589AMXT were planted in 30-inch row spacing at a rate of 2 seeds/foot on May 22. All fungicide applications were applied at 15 gal/acre and 40 psi using a Lee self-propelled sprayer equipped with a 10-foot boom, fitted with six TJ-VS 8002 nozzles spaced 20 inches apart. Delaro Complete fungicide was applied on July 12, July 31, August 16, and September 5 at the 10-leaf (V10), tassel/silk (VT/R1), blister (R2), and dough (R4) growth stages, respectively. A weather-based prediction model, Tarspotter, was used, and applications were made on August 22 and September 4 at the R2 + 6 days and R4 growth stages, respectively. Disease ratings were assessed on August 26, September 13, and September 25 at milk (R3), dent (R5), and maturity (R6) growth stages, respectively. Tar spot was rated by visually assessing the percentage of stromata (0–100%) per leaf on five plants in each plot at the ear leaf. Values for each plot were averaged before analysis. The two center rows of each plot were harvested on October 18, and yields were adjusted to 15.5% moisture. All data were analyzed in SAS 9.4 (SAS Institute, Cary, NC). A generalized linear mixed model analysis of variance was performed using PROC GLIMMIX. Values are least squares means, and values with different letters are significantly different based on Fisher's least significant difference ($\alpha = 0.05$).

In 2024, weather conditions were favorable for tar spot. Tar spot was first detected in the field on the lower canopy on July 15. Tar spot was the most prominent disease in the trial, with low levels of northern corn leaf blight present. There was a significant interaction between hybrid and fungicide for disease, but data for main effects of hybrid and treatment are presented for brevity (Table 19). The tar spot susceptible hybrid (W2585VT2P) had significantly more tar spot on August 26, September 13, and September 25 as compared to the tar spot resistant hybrid (P0589AMXT). On September 26, applications of Delaro Compete at V10 and VT/R1 and Tarspotter significantly reduced tar spot stromata severity compared to the nontreated control. On September 13 and September 25, all timings of Delaro Complete significantly reduced tar spot over nontreated control. By September 25, applications of Delaro Complete at VT/R1 had the lowest level of tar spot, followed by applications at R2 and applications made by Tarspotter as compared to V10 and R4 timings. No significant differences between treatments and nontreated control were observed for grain moisture and test weight. No significant differences between hybrids for grain yield were detected, but fungicide applications at VT/R1, V10, and R2 significantly increased yield over the nontreated control.

TABLE 19. *Effect of Fungicide on Tar Spot Severity and Yield of Corn*

TREATMENT, RATE/ACRE, AND TIMING[z]	TAR SPOT %[Y] AUGUST 26	TAR SPOT %[Y] SEPTEMBER 13	TAR SPOT %[Y] SEPTEMBER 25	HARVEST MOISTURE %	TEST WEIGHT LB/BU	YIELD[x] BU/ACRE
Hybrid						
W2583VT2P	0.49 a	6.0 a	14.7 a	17.2	55.0 b	219.5
Treatment						
Po589AMXT	0.22 b	1.7 b	9.2 b	17.7	55.7 a	216.2
Nontreated control	0.52 a	8.5 a	21.4 a	17.0	55.4	209.3 c
Delaro Complete 3.82 SC 8.0 fl oz at V10	0.22 c	3.6 b	14.9 b	17.1	55.7	221.8 ab
Delaro Complete 3.82 SC 8.0 fl oz at VT/R1	0.17 c	1.1 c	5.8 e	17.7	55.4	225.5 a
Delaro Complete 3.82 SC 8.0 fl oz at R2	0.41 ab	3.3 bc	9.3 d	17.2	55.3	219.2 ab
Delaro Complete 3.82 SC 8.0 fl oz at R4	0.44 ab	4.1 b	12.0 c	17.5	55.4	214.5 bc
Delaro Complete 3.82 SC 8.0 fl oz at Tarspotter	0.38 b	2.6 bc	8.4 d	18.1	54.9	216.7 abc
P-value hybrid[w]	0.0001	<.0001	<.0001	0.2889	0.0074	0.2410
P-value treatment	0.0001	<.0001	<.0001	0.5811	0.6500	0.0294
P-value hybrid by treatment	0.2518	0.0044	0.5842	0.8969	0.2656	0.5464

[z] Fungicide treatments were applied on July 12, July 31, August 16, and September 4 at the V10 (10-leaf), tassel/silk (VT/R1), blister (R2), and dough (R4) growth stages, respectively. Tarspotter applications were made on August 22 and September 4 at R2 + 6 days fb R4 growth stage. fb=followed by.

[y] Tar spot stroma were visually assessed as a percentage (0–100%) of leaf area on five plants in each plot at the ear leaf (EL).

[x] Yields were adjusted to 15.5% moisture and harvested on October 18.

[w] All data were analyzed in SAS 9.4 (SAS Institute, Cary, NC). A generalized linear mixed model analysis of variance was performed using PROC GLIMMIX. Values are least squares means, and values with different letters are significantly different based on Fisher's least significant difference (α = 0.05).

EVALUATION TILLAGE, HYBRID, AND FUNGICIDE FOR FOLIAR DISEASES IN CORN IN NORTHWESTERN INDIANA, 2024 (COR24-05.PPAC)

M. Acevedo, S. Shim, and D. E. P. Telenko, Department of Botany and Plant Pathology, Purdue University, West Lafayette, IN 47907-2054

CORN (*ZEA MAYS* W2585SSR AND P0589AMXT)

Tar spot, *Phyllachora maydis*
Gray leaf spot, *Cercospora zeae-maydis*
Northern corn leaf blight, *Exserohilum turcicum*

A trial was established at the Pinney Purdue Agricultural Center (PPAC) in Porter County, Indiana. The experiment was a split plot with six replications. Plots were 10 feet wide and 30 feet long and consisted of four rows, and the two center rows were used for evaluation. The previous crops were corn in the no-till block and soybean in the tilled block. Standard practices for grain corn production in Indiana were followed. The tillage blocks (no-till and full-tillage) was the main effect. Two corn hybrids and fungicide application (yes/no) were factorial arrangement in the subplots. Corn hybrids W2585VT2PRIB (susceptible) and P0589AMXT (moderately resistant) were planted in 30-inch row spacing at a rate of 2 seeds/foot on May 22. A fungicide (Veltyma at 7.0 fl oz/acre) was applied on August 7 at silk (R1) growth stage. The fungicide was applied at 15 gal/acre and 40 psi using a Lee self-propelled sprayer equipped with a 10-foot boom, fitted with six TJ-VS 8002 nozzles spaced 20 inches apart. Tar spot was rated by visually assessing the percentage of stroma per leaf on 10 plants in each plot at the ear leaf on September 26 at dent/maturity (R5/R6) growth stage. Gray leaf spot (GLS) and northern corn leaf blight (NCLB) were rated by visually assessing the percentage severity on ear leaf on 10 plants on September 13 at dough/dent (R4/R5) growth stage. Values for each plot were averaged before analysis. The two center rows of each plot were harvested on October 22, and yields were adjusted to 15.5% moisture. All data were analyzed in SAS 9.4 (SAS Institute, Cary, NC). A generalized linear mixed model analysis of variance was performed using PROC GLIMMIX. Values are least squares means, and values with different letters are significantly different based on Fisher's least significant difference ($\alpha = 0.05$).

In 2024, weather conditions were favorable for foliar diseases. Tar spot, GLS, and NCLB were present in the plots. Tar spot was the most prominent disease in the trial and reached moderate severity. There was a significant interaction between hybrid and Veltyma application for tar spot severity; therefore, data are presented for that interaction (Table 20). No significant difference was detected between no-tillage and tillage for tar spot severity. GLS and NCLB were significantly higher in the tillage verses no-till plots. Harvest moisture was reduced under tillage versus no tillage, while test weight and grain yield were highest in the tillage versus no-tillage treatments. Tar spot severity was highest in the susceptible hybrid with no fungicide application. Veltyma significantly reduced tar spot severity compared to nontreated in both hybrids. There were no significant differences between hybrid and fungicide programs for GLS and NCLB severity, harvest moisture, test weight, and yield of corn.

TABLE 20. *Effect of Tillage, Hybrid, and Fungicide on Foliar Disease and Yield of Corn*

TILLAGE, HYBRID, TREATMENT, AND RATE/ACRE[z]	TAR SPOT % STROMATA[y]	GLS % SEVERITY[x]	NCLB % SEVERITY[x]	MOISTURE %	TEST WEIGHT LB/BU	YIELD[w] BU/ACRE
Tillage						
No-tillage (high residue)	8.2	0.1 b	0.0 b	18.4 a	55.9 b	189.5 b
Yes-tillage (low residue)	9.5	0.4 a	3.3 a	15.5 b	57.9 a	211.3 a
Hybrid and treatment						
Susceptible; Nontreated control	24.9 a	0.4	3.0	17.1	56.0	189.7
Susceptible; Veltyma 3.34 SC 7.0 fl oz	2.4 c	0.1	1.7	16.8	56.8	209.0
Moderately resistant; Nontreated control	6.9 b	0.4	1.7	16.5	57.3	194.5
Moderately resistant; Veltyma 3.34 SC 7.0 fl oz	1.1 c	0.1	0.4	17.6	57.5	208.5
P-value tillage[v]	0.2642	0.0094	0.0001	0.0001	0.0001	0.0001
P-value hybrid	0.0001	0.9481	0.0075	0.8255	0.0019	0.5363
P-value fungicide	0.0001	0.0108	0.0097	0.4080	0.1030	0.0001
P-value tillage*hybrid	0.2140	0.3063	0.0048	0.7731	0.0373	0.5433
P-value tillage*fungicide	0.0996	0.0940	0.0150	0.3712	0.2139	0.6046
P-value variety*fungicide	0.0001	0.7816	0.9993	0.1552	0.3017	0.4487
P-value tillage*hybrid*fungicide	0.4668	0.1775	0.8626	0.8520	0.3143	0.4106

[z] Veltyma application was applied at 15 gal/acre and 40 psi using a Lee self-propelled sprayer equipped with a 10-foot boom, fitted with six TJ-VS 8002 nozzles spaced 20 inches apart. Veltyma was applied on August 7 at silk growth stage (R1).

[y] Tar spot stroma was visually assessed as a percentage (0–100%) of ear leaf on 10 plants in each plot on September 26 at dent/maturity (R5/R6) growth stage.

[x] GLS and NCLB severity visually assessed as a percentage (0–100%) of leaf area on 10 plants in each plot on September 13 at dough/dent (R4/R5) growth stage. GLS = gray leaf spot; NCLB = northern corn leaf blight.

[w] Yields were adjusted to 15.5% moisture and harvested on October 22.

[v] All data were analyzed in SAS 9.4 (SAS Institute, Cary, NC). A generalized linear mixed model analysis of variance was performed using PROC GLIMMIX. Values are least squares means, and values with different letters are significantly different based on Fisher's least significant difference (α = 0.05).

EVALUATION OF PLANTING DATE, HYBRID, AND FUNGICIDE FOR TAR SPOT IN CORN IN NORTHWESTERN INDIANA, 2024 (COR24-06.PPAC)

I. L. Miranda, S. Shim, and D. E. P. Telenko, Department of Botany and Plant Pathology, Purdue University, West Lafayette, IN 47907-2054

CORN (*ZEA MAY* P9608Q AND P1108Q)

Tar spot, *Phyllachora maydis*

A trial was established at the Pinney Purdue Agricultural Center (PPAC) in Porter County, Indiana. The experiment was a randomized complete block design with four replications. Plots were 10 feet wide and 30 feet long and consisted of four rows, and the two center rows were used for evaluation. The previous crop was corn. Standard practices for nonirrigated grain corn production in Indiana were followed. Corn hybrids P9608Q and P1108Q were planted in 30-inch row spacing at a rate of 2 seeds/foot on April 25 and May 23. Foliar applications were made at the tassel/silk (VT/R1) growth stage on July 12 for April 25 planting and July 30 for May 23 planting plots. Foliar fungicide applications were applied using a CO_2 backpack sprayer at 5 mph for the first planting date (April 25) and at 15 gal/acre at 40 psi using a Lee self-propelled sprayer equipped with a 10-foot boom, fitted with six TJ-VS 8002 nozzles spaced 20 inches apart, at 3 mph for the second planting date (May 23). Disease ratings were assessed on September 6, September 13, and September 23 at dent (R5) and physiological maturity (R6) growth stages. Tar spot stromata severity visually was assessed as a percentage (0–100%) of symptomatic leaf area at ear leaf on five plants per plot and averaged before analysis. The two center rows of each plot were harvested on October 21, and yields were adjusted to 15.5% moisture. All data were analyzed in SAS 9.4 (SAS Institute, Cary, NC). A generalized linear mixed model analysis of variance was performed using PROC GLIMMIX. Values are least squares means, and values with different letters are significantly different based on Fisher's least significant difference (α = 0.05).

In 2024, weather conditions were favorable for disease. Tar spot was the most prominent disease in the trial and reached moderate severity. On September 6, tar spot severity was significantly reduced by Veltyma compared to the nontreated plants, while no significant differences were found between plantings and hybrids (Table 21). On September 13, in the April 25 planting, the Veltyma application was not significantly different from the nontreated control in both hybrids, but in the May 23 planting Veltyma still significantly reduced tar spot severity over the nontreated controls in both hybrids. On September 23, tar spot severity was less than 10% in both hybrids planted on May 23 that included a Veltyma treatment. Hybrid P1108Q planted on May 23 had significantly higher moisture compared to April 25 planting and the P9608Q hybrid. P1108Q had significantly higher yield compared to hybrid P9608Q when planted on April 25. No differences between hybrids and treatments were detected for yield when planted on May 23.

TABLE 21. *Effect of Planting Date, Hybrid and Fungicide on Tar Spot Severity, Canopy Greenness, Harvest Moisture, and Yield of Corn*

PLANTING DATE, HYBRID, AND TREATMENT RATE/ACRE[z]	TAR SPOT % SEVERITY[y] SEPTEMBER 6	TAR SPOT % SEVERITY[y] SEPTEMBER 13	TAR SPOT % SEVERITY[y] SEPTEMBER 23	HARVEST MOISTURE %	YIELD[x] BU/ACRE
Planting date A, P9608Q, Nontreated control	9.4 ab	16.8 bcd	22.5 ab	13.9 b	170.0 d
Planting date A, P9608Q, Veltyma 7.0 fl oz	3.9 c	12.3 d	16.2 c	14.6 b	188.1 cd
Planting date A, P1099Q, Nontreated control	8.4 b	21.0 abc	18.6 bc	14.5 b	224.6 ab
Planting date A, P1099Q, Veltyma 7.0 fl oz	4.4 c	16.0 cd	21.9 ab	14.8 b	235.7 a
Planting date B, P9608Q, Nontreated control	12.9 a	25.5 a	24.4 a	14.7 b	202.4 bc
Planting date B, P9608Q, Veltyma 7.0 fl oz	4.0 c	5.5 e	8.5 d	15.5 b	208.6 bc
Planting date B, P1099Q, Nontreated control	10.1 ab	21.8 ab	23.8 ab	18.2 a	203.4 bc
Planting date B, P1099Q, Veltyma 7.0 fl oz	2.2 c	4.1 e	7.0 d	18.1 a	218.4 ab
P-value[w]	0.0001	0.0001	0.0001	0.0003	0.0015

[z] Foliar applications were made at the tassel/silk (VT/R1) growth stage on July 12 for the April 25 planting and July 30 for the May 23 planting.

[y] Tar spot stromata was visually assessed as a percentage (0–100%) of leaf area on five plants in each plot at the ear leaf (EL) on September 6, September 13, and September 23.

[x] Yields were adjusted to 15.5% moisture and harvested on October 21.

[w] All data were analyzed in SAS 9.4 (SAS Institute, Cary, NC). A generalized linear mixed model analysis of variance was performed using PROC GLIMMIX. Values are least squares means, and values with different letters are significantly different based on Fisher's least significant difference (α = 0.05).

EVALUATION OF FUNGICIDES FOR TAR SPOT IN CORN IN NORTHWESTERN INDIANA, 2024 (CORN 24-07.PPAC)

S. Shim and D. E. P. Telenko, Department of Botany and Plant Pathology, Purdue University, West Lafayette, IN 47907-2054

CORN (*ZEA MAYS* W2584 VT2P RIB)

Tar spot, *Phyllachora maydis*

A trial was established at the Pinney Purdue Agricultural Center (PPAC) in Porter County, Indiana. The experiment was a randomized complete block design with four replications. Plots were 10 feet wide and 30 feet long and consisted of four rows, and the two center rows were used for evaluation. The previous crop was corn. Standard practices for irrigated grain corn production in Indiana were followed. Corn hybrid W2584 VT2P RIB was planted in 30-inch row spacing at a rate of 34,000 seeds/acre on May 23. The field was overhead irrigated weekly at 1 inch unless weekly rainfall was 1 inch or higher to encourage disease. All products were applied at 15 gal/acre and 40 psi using a Lee self-propelled sprayer equipped with a 10-foot boom, fitted with six TJ-VS 8002 nozzles spaced 20 inches apart. Treatments were applied on August 7 at silk (R1) growth stage. Disease ratings were assessed on September 4 at dough/early dent (R4/R5) and on September 13 at dent (R5.30) growth stages. Tar spot stromata severity was visually assessed as a percentage (0–100%) of symptomatic leaf area at ear leaf on five plants per plot and averaged before analysis. Values for the five leaves were averaged before analysis. The two center rows of each plot were harvested on October 23, and yields were adjusted to 15.5% moisture. All data were analyzed in SAS 9.4 (SAS Institute, Cary, NC). A generalized linear mixed model analysis of variance was performed using PROC GLIMMIX. Values are least squares means, and values with different letters are significantly different based on Fisher's least significant difference ($\alpha = 0.05$).

In 2024, weather conditions were moderately favorable for the disease. Tar spot was first detected in the field on July 19. All treatments significantly reduced tar spot over nontreated controls September 4 and September 13 (Table 22). No significant differences were detected for canopy greenness, harvest moisture, test weight, and yield of corn.

TABLE 22. *Effect of Fungicide on Tar Spot Severity, Canopy Greenness, and Yield of Corn*

TREATMENT AND RATE/ACRE[z]	TAR SPOT % STROMATA[y] SEPTEMBER 4	TAR SPOT % STROMATA[y] SEPTEMBER 13	CANOPY GREEN[x] %	HARVEST MOISTURE %	TEST WEIGHT LB/BU	YIELD[w] BU/ACRE
Nontreated control	1.4 a	14.1 a	63.8	18.0	58.0	175.3
Maxentis SC 4.8 fl oz	0.4 de	3.3 b	75.0	17.5	57.7	188.2
Maxentis SC 9.6 fl oz	0.2 e	3.1 b	71.3	18.7	58.1	191.2
Stratego YLD 5.0 fl oz	0.7 cd	3.6 b	72.5	19.7	57.5	183.9
Veltyma 3.34 SC 10.0 fl oz	1.0 b	5.0 b	75.0	17.5	58.6	199.3
Headline AMP 14.4 fl oz	0.9 bc	4.2 b	73.8	19.1	57.5	183.0
P-value[v]	0.0001	0.0001	0.0335	0.5086	0.6821	0.4027

[z] Fungicides were applied on August 7 at silk (R1) growth stage.

[y] Tar spot stromata was visually assessed as a percentage (0–100%) of leaf area on five plants in each plot at the ear leaf on September 4 and September 13 at dough/early dent (R4/R5) and dent (R5.30) growth stages, respectively.

[x] Canopy greenness as %, visually rated per plot, was assessed on September 13 at dent (R5) growth stage.

[w] Yields were adjusted to 15.5% moisture and harvested on October 23.

[v] All data were analyzed in SAS 9.4 (SAS Institute, Cary, NC). A generalized linear mixed model analysis of variance was performed using PROC GLIMMIX. Values are least squares means, and values with different letters are significantly different based on Fisher's least significant difference (α = 0.05).

EVALUATION OF FUNGICIDES FOR TAR SPOT IN CORN IN NORTHWESTERN INDIANA, 2024 (CORN 24-14.PPAC)

S. Shim and D. E. P. Telenko, Department of Botany and Plant Pathology,
Purdue University, West Lafayette, IN 47907-2054

CORN (*ZEA MAYS* W2584 VT2P RIB)

Tar spot, *Phyllachora maydis*

A trial was established at the Pinney Purdue Agricultural Center (PPAC) in Porter County, Indiana. The experiment was a randomized complete block design with four replications. Plots were 10 feet wide and 30 feet long and consisted of four rows, and the two center rows were used for evaluation. The previous crop was corn. Standard practices for nonirrigated grain corn production in Indiana were followed. Corn hybrid W2584 VT2P RIB was planted in 30-inch row spacing at a rate of 2 seeds/foot on May 23. The 2x2 treatments were applied at planting in 10 gal/acre. Foliar fungicides were applied at 15 gal/acre and 40 psi using a CO_2 backpack sprayer at V10 and a Lee self-propelled sprayer equipped with a 10-foot boom, fitted with six TJ-VS 8002 nozzles spaced 20 inches apart, at VT/R1. Fungicides were applied on July 12 at V10 and July 31 at tassel/silk (VT/R1) growth stages. Disease ratings were assessed on September 4 at dough (R4) and September 13 at dent (R5) growth stages. Tar spot stromata severity was visually assessed as a percentage (0–100%) of symptomatic leaf area at ear leaf on five plants per plot and averaged before analysis. Values for the five leaves were averaged before analysis. The two center rows of each plot were harvested on October 21, and yields were adjusted to 15.5% moisture. All data were analyzed in SAS 9.4 (SAS Institute, Cary, NC). A generalized linear mixed model analysis of variance was performed using PROC GLIMMIX. Values are least squares means, and values with different letters are significantly different based on Fisher's least significant difference ($\alpha = 0.05$).

In 2024, weather conditions were moderately favorable for the disease. Tar spot was the most prominent disease in the trial. Tar spot was first detected in plots on August 2. No significant differences between treatments were detected for tar spot severity on September 4 (Table 23). All fungicide programs reduced tar spot significantly compared to nontreated control on September 13. No significant differences were detected for canopy greenness, harvest moisture, test weight, and yield of corn.

TABLE 23. *Effect of Treatment on Tar Spot Severity and Yield of Corn*

TREATMENT, RATE/ACRE, AND TIMING[z]	TAR SPOT % SEVERITY[y] SEPTEMBER 4	TAR SPOT % SEVERITY[y] SEPTEMBER 13	HARVEST MOISTURE %	TEST WEIGHT LB/BU	YIELD[x] BU/ACRE
Nontreated control	0.10	9.1 a	15.7	57.4	209.8
Veltyma 3.34 SC 7.0 fl oz + NIS at 0.25% v/v at VT/R1	0.06	1.1 f	16.1	57.6	209.5
Delaro Complete 8.0 fl oz + NIS at 0.25% v/v at VT/R1	0.06	1.2 ef	17.0	57.0	207.4
Topguard 8.0 fl oz at V10	0.08	5.5 bc	17.0	56.8	210.7
Adastrio 8.0 fl oz at V10	0.07	5.6 b	16.6	57.6	203.5
Xyway LFR 9.5 fl oz 2x2 at plant fb Adastrio 7.0 fl oz + NIS at 0.25% v/v at VT/R1	0.05	2.4 def	18.0	56.7	202.0
Xyway LFR 15.2 fl oz 2x2 at plant fb Veltyma 3,34 SC 7.0 fl oz + NIS at 0.25% v/v at VT/R1	0.07	0.6 f	16.3	57.3	217.8
Xyway LFR 15.2 fl oz 2x2 at plant	0.13	6.4 b	16.8	57.4	201.7
Veltyma 3.34 SC 7.0 fl oz + OR-099EPA 0.4% v/v at VT/R1	0.04	0.8 f	16.5	57.5	214.3
Delaro Complete 8.0 fl oz + OR-099EPA 0.4% v/v at VT/R1	0.10	1.8 def	16.6	57.4	212.6
Cortina Xtra 8.0 fl oz at VT/R1	0.04	1.7 def	17.0	57.0	211.5
Cortina Xtra 8.0 fl oz + Nutex EDA 8.0 fl oz at VT/R1	0.04	1.7 def	16.5	57.3	207.8
Cortina Xtra 12.0 fl oz at VT/R1	0.05	3.1 de	18.3	56.5	211.4
SA-0050010 5.7 fl oz at VT/R1	0.05	3.4 d	16.5	57.3	210.3
SA-0050010 5.7 fl oz + Nutex EDA 8.0 fl oz at VT/R1	0.09	3.5 cd	17.2	56.7	211.2
Maxentis SC 9.6 fl oz at VT/R1	0.05	1.9 def	16.7	57.4	213.0
P-value[w]	*0.1036*	*0.0001*	*0.3832*	*0.6380*	*0.7033*

[z] The 2x2 treatments were applied at planting in 10 gal/acre. Foliar fungicides were applied on July 12 at V10 and July 31 at tassel/silk (VT/R1) growth stages in 15 gal/acre. fb = followed by.

[y] Tar spot stromata was visually assessed as a percentage (0–100%) of leaf area on five plants in each plot at the ear leaf on September 4 at dough (R4) and September 13 at dent (R5.40) growth stages.

[x] Yields were adjusted to 15.5% moisture and harvested on October 21.

[w] All data were analyzed in SAS 9.4 (SAS Institute, Cary, NC). A generalized linear mixed model analysis of variance was performed using PROC GLIMMIX. Values are least squares means, and values with different letters are significantly different based on Fisher's least significant difference ($\alpha = 0.05$).

INDUSTRY SPONSORED FUNGICIDE EVALUATION FOR TAR SPOT IN CORN IN NORTHWESTERN INDIANA, 2024 (CORN 24-18.PPAC)

S. Shim and D. E. P. Telenko, Department of Botany and Plant Pathology, Purdue University, West Lafayette, IN 47907-2054

CORN (*ZEA MAYS* W2584 VT2P RIB)

Tar spot, *Phyllachora maydis*

A trial was established at the Pinney Purdue Agricultural Center (PPAC) in Porter County, Indiana. The experiment was a randomized complete block design with four replications. Plots were 10 feet wide and 30 feet long and consisted of four rows, and the two center rows were used for evaluation. The previous crop was corn. Standard practices for nonirrigated grain corn production in Indiana were followed. Corn hybrid W2584 VT2P RIB was planted in 30-inch row spacing at a rate of 34,000 seeds/acre on May 23. The field was overhead irrigated weekly at 1 inch unless weekly rainfall was 1 inch or higher to encourage disease. All fungicides were applied at 15 gal/acre and 40 psi using a Lee self-propelled sprayer equipped with a 10-foot boom, fitted with six TJ-VS 8002 nozzles spaced 20 inches apart. Fungicides were applied on August 7 at silk (R1) and August 20 milk (R3) growth stages and included NIS (Masterlock 6.4 fl oz/acre). Disease ratings were assessed on September 5 at beginning dent (R5) and September 12 at dent (R5.40) growth stages. Tar spot stromata severity was visually assessed as a percentage (0–100%) of symptomatic leaf area at ear leaf on five plants per plot and averaged before analysis. Values for the five leaves were averaged before analysis. The two center rows of each plot were harvested on October 22, and yields were adjusted to 15.5% moisture. All disease and yield data were analyzed using a mixed model analysis of variance, and means were separated using Fisher's least significant difference ($\alpha = 0.05$).

In 2024, weather conditions were moderately favorable for the diseases. Tar spot was first detected in plots on July 19. All fungicide treatments significantly reduced tar spot as compared to nontreated control on September 5 and September 12 (Table 24). On September 5, tar spot has the lowest severity with Miravis Neo at R1 followed by Miravis Neo at R3, but this was not significantly different from Trivapro or Miravis Neo applied at R1. On September 12 the Miravis Neo at R1 followed by Miravis Neo at R3 had the lowest level of tar spot as compared to the other programs. No significant differences were detected for canopy greenness, harvest moisture, and test weight. All fungicides programs increased grain yield over the nontreated control.

TABLE 24. *Effect of Fungicide Treatment on Tar Spot Severity, Canopy Greenness, and Yield of Corn*

TREATMENT, RATE/ACRE, AND TIMING[z]	TAR SPOT % STROMATA[y] SEPTEMBER 5	TAR SPOT % STROMATA[y] SEPTEMBER 12	CANOPY GREEN[x] %	HARVEST MOISTURE %	TEST WEIGHT LB/BU	YIELD[w] BU/ACRE
Nontreated control	7.4 a	26.0 a	76.3	15.3	57.8	178.7 c
Trivapro 13.7 fl oz at R1	2.0 de	9.1 cd	78.8	18.1	56.1	190.8 b
Miravis Neo 13.7 fl oz at R1	2.2 de	9.2 cd	75.0	17.6	57.5	193.4 ab
Veltyma 7.0 fl oz at R1	2.8 cd	9.8 cd	78.8	16.8	57.8	195.5 ab
Delaro Complete 8.0 fl oz at R1	2.7 d	6.8 de	76.3	16.5	58.2	200.5 a
Adastrio 7.0 fl oz at R1	2.5 d	9.5 cd	73.8	18.6	56.9	189.1 b
Miravis Neo 13.7 fl oz at R3	3.9 bc	15.4 b	72.5	17.5	57.2	195.3 ab
Miravis Neo 13.7 fl oz at R1 fb Miravis Neo 13.7 fl oz at R3	1.3 e	4.5 e	78.8	17.2	58.2	198.1 ab
P-value[v]	*0.0001*	*0.0001*	*0.8427*	*0.0698*	*0.1975*	*0.0056*

[z] Fungicides were applied on July 30 at silk (R1) growth stage and included NIS (Masterlock 6.4 fl oz). fb = followed by.

[y] Tar spot stromata was visually assessed as a percentage (0–100%) of leaf area on five plants in each plot at the ear leaf on September 5 at beginning dent (R5) and September 12 at dent (R5.40) growth stages.

[x] Canopy greenness as %, visually rated per plot, was assessed on September 12 at dent (R5) growth stage.

[w] Yields were adjusted to 15.5% moisture and harvested on October 22.

[v] All data were analyzed in SAS 9.4 (SAS Institute, Cary, NC). A generalized linear mixed model analysis of variance was performed using PROC GLIMMIX. Values are least squares means, and values with different letters are significantly different based on Fisher's least significant difference test (α = 0.05).

EVALUATION OF FUNGICIDES FOR TAR SPOT IN CORN IN NORTHWESTERN INDIANA, 2024 (CORN 24-22.PPAC)

S. Shim and D. E. P. Telenko, Department of Botany and Plant Pathology,
Purdue University, West Lafayette, IN 47907-2054

CORN (*ZEA MAYS* W2584 VT2P RIB)

Tar spot, *Phyllachora maydis*

A trial was established at the Pinney Purdue Agricultural Center (PPAC) in Porter County, Indiana. The experiment was a randomized complete block design with four replications. Plots were 10 feet wide and 30 feet long and consisted of four rows, and the two center rows were used for evaluation. The previous crop was corn. Standard practices for irrigated grain corn production in Indiana were followed. Corn hybrid W2584 VT2P RIB was planted in 30-inch row spacing at a rate of 34,000 seeds/acre on May 23. The field was overhead irrigated weekly at 1 inch unless weekly rainfall was 1 inch or higher to encourage disease. All fungicides were applied at 15 gal/acre and 40 psi using a Lee self-propelled sprayer equipped with a 10-foot boom, fitted with six TJ-VS 8002 nozzles spaced 20 inches apart. Fungicides were applied on June 24 at V5 and August 7 at silk (R1) growth stages. Disease ratings were assessed on September 4 at dough (R4) and September 13 at dent (R5.30) growth stages. Tar spot stromata severity was visually assessed as a percentage (0–100%) of symptomatic leaf area at ear leaf on five plants per plot and averaged before analysis. Values for the five leaves were averaged before analysis. The two center rows of each plot were harvested on October 23, and yields were adjusted to 15.5% moisture. All data were analyzed in SAS 9.4 (SAS Institute, Cary, NC). A generalized linear mixed model analysis of variance was performed using PROC GLIMMIX. Values are least squares means, and values with different letters are significantly different based on Fisher's least significant difference (α = 0.05).

In 2024, weather conditions were moderately favorable for the diseases. Tar spot was the most prominent disease and was first detected in the field on July 19. All fungicide treatments significantly reduced tar spot as compared to nontreated control on September 4 and September 13, except Affiance applied at V5 on September 13 (Table 25). On September 5 and September 13, tar spot had the lowest severity with Miravis Neo applied at R1, but this was not significantly different from Delaro Complete or Affiance applied at R1. All fungicide programs increased canopy greenness on September 13 over nontreated control. No significant differences were detected for moisture and test weight of corn. All fungicides programs increased grain yield over nontreated control except Affiance at V5. Affiance at V5 followed by Veltyma at R1 and Veltyma at R1 had the highest yield but were not significantly different from Delaro Complete at R1 and Miravis Neo at R1.

TABLE 25. *Effect of Treatment on Tar Spot Severity, Canopy Greenness, and Yield of Corn*

TREATMENT, RATE/ACRE, AND TIMING[z]	TAR SPOT % STROMATA[y] SEPTEMBER 4	TAR SPOT % STROMATA[y] SEPTEMBER 13	CANOPY GREEN[x] %	HARVEST MOISTURE %	TEST WEIGHT LB/BU	YIELD[w] BU/ACRE
Nontreated control	1.2 a	13.6 a	63.8 c	17.2	57.7	176.1 c
Affiance 10.0 fl oz at V5	0.8 b	14.5 a	70.0 b	18.2	57.4	173.0 c
Affiance 10.0 fl oz at R1	0.3 cd	3.5 bc	77.5 a	19.4	57.0	180.4 bc
Affiance 10.0 fl oz at V5 fb Veltyma 7.8 fl oz at R1	0.6 b	5.0 b	77.5 a	19.2	57.5	199.5 a
Veltyma 7.8 fl oz at R1	0.6 b	5.3 b	78.8 a	20.4	56.7	199.5 a
Delaro Complete 7.5 fl oz at R1	0.3 c	2.6 bc	78.8 a	19.9	57.0	191.8 ab
Miravis Neo 13.6 fl oz at R1	0.1 d	1.6 c	78.8 a	17.2	58.1	193.8 ab
P-value[v]	*0.0001*	*0.0001*	*0.0001*	*0.1987*	*0.7943*	*0.0017*

[z] Fungicides were applied on June 24 at V5 and August 7 at silk (R1) growth stages. fb = followed by.

[y] Tar spot stromata was visually assessed as a percentage (0–100%) of leaf area on five plants in each plot at the ear leaf on September 4 at dough (R4) and September 13 at dent (R5.30) growth stages.

[x] Canopy greenness as % was visually rated per plot and was assessed on September 13.

[w] Yields were adjusted to 15.5% moisture and harvested on October 23.

[v] All data were analyzed in SAS 9.4 (SAS Institute, Cary, NC). A generalized linear mixed model analysis of variance was performed using PROC GLIMMIX. Values are least squares means, and values with different letters are significantly different based on Fisher's least significant difference (α = 0.05).

EVALUATION OF FUNGICIDES FOR TAR SPOT IN CORN IN NORTHWESTERN INDIANA, 2024 (CORN 24-28.PPAC)

S. Shim and D. E. P. Telenko, Department of Botany and Plant Pathology,
Purdue University, West Lafayette, IN 47907-2054

CORN (*ZEA MAYS* W2584 VT2P RIB)

Tar spot, *Phyllachora maydis*

A trial was established at the Pinney Purdue Agricultural Center (PPAC) in Porter County, Indiana. The experiment was a randomized complete block design with four replications. Plots were 10 feet wide and 30 feet long and consisted of four rows, and the two center rows were used for evaluation. The previous crop was corn. Standard practices for nonirrigated grain corn production in Indiana were followed. Corn hybrid W2584 VT2P RIB was planted in 30-inch row spacing at a rate of 34,000 seeds/acre on May 23. All fungicides were applied at 15 gal/acre and 40 psi using a Lee self-propelled sprayer equipped with a 10-foot boom, fitted with six TJ-VS 8002 nozzles spaced 20 inches apart. Fungicides were applied on July 30 at silk (R1) growth stage. Disease ratings were assessed on August 30 at milk (R3) and September 12 at dent (R5.40) growth stages. Tar spot stromata severity was visually assessed as a percentage (0–100%) of symptomatic leaf area at ear leaf on five plants per plot and averaged before analysis. Values for the five leaves were averaged before analysis. The two center rows of each plot were harvested on October 21, and yields were adjusted to 15.5% moisture. All disease and yield data were analyzed in SAS 9.4 (SAS Institute, Cary, NC). A generalized linear mixed model analysis of variance was performed using PROC GLIMMIX. Values are least squares means, and values with different letters are significantly different based on Fisher's least significant difference (α = 0.05).

In 2024, weather conditions were moderately favorable for the diseases. Tar spot was the most prominent disease and was first detected in plots on August 5. All fungicide treatments significantly reduced tar spot as compared to nontreated control on August 30 and September 12 (Table 26). No significant differences were detected for canopy greenness, harvest moisture, test weight, and yield of corn.

TABLE 26. *Effect of Treatment on Tar Spot Severity, Canopy Greenness, and Yield of Corn*

TREATMENT AND RATE/ACRE[z]	TAR SPOT % STROMATA[y] AUGUST 30	TAR SPOT % STROMATA[y] SEPTEMBER 12	CANOPY GREEN[x] %	HARVEST MOISTURE %	TEST WEIGHT LB/BU	YIELD[w] BU/ACRE
Nontreated control	0.6 a	10.1 a	70.0	16.2	56.9	203.2
Veltyma 7.0 fl oz	0.2 b	1.6 b	75.0	17.7	57.4	215.1
Miravis Neo 13.7 fl oz	0.1 b	3.3 b	78.8	18.6	64.6	205.7
Delaro Complete 8.0 fl oz	0.2 b	3.4 b	73.8	16.2	56.8	213.6
Adastrio 8.0 fl oz	0.1 b	3.3 b	77.5	16.9	55.8	209.2
P-value[v]	*0.0001*	*0.0001*	*0.0669*	*0.1762*	*0.5679*	*0.6637*

[z] Fungicides were applied on July 30 at silk (R1) growth stage and included NIS (Preference) 0.25% v/v.

[y] Tar spot stromata was visually assessed as a percentage (0–100%) of leaf area on five plants in each plot at the ear leaf on August 30 at milk (R3) and September 12 at dent (R5.40) growth stages.

[x] Canopy greenness as a percentage (0–100%) was visually rated on September 12.

[w] Yields were adjusted to 15.5% moisture and harvested on October 21.

[v] All data were analyzed in SAS 9.4 (SAS Institute, Cary, NC). A generalized linear mixed model analysis of variance was performed using PROC GLIMMIX. Values are least squares means, and values with different letters are significantly different based on Fisher's least significant difference ($\alpha = 0.05$).

EVALUATION OF FUNGICIDE FOR TAR SPOT IN CORN IN NORTHWESTERN INDIANA, 2024 (CORN 24-29.PPAC)

S. Shim and D. E. P. Telenko, Department of Botany and Plant Pathology, Purdue University, West Lafayette, IN 47907-2054

CORN (*ZEA MAYS* W2584 VT2P RIB)

Tar spot, *Phyllachora maydis*

A trial was established at the Pinney Purdue Agricultural Center (PPAC) in Porter County, Indiana. The experiment was a randomized complete block design with four replications. Plots were 10 feet wide and 30 feet long and consisted of four rows, and the two center rows were used for evaluation. The previous crop was corn. Standard practices for nonirrigated grain corn production in Indiana were followed. Corn hybrid W2584 VT2P RIB was planted in 30-inch row spacing at a rate of 34,000 seeds/acre on May 23. All fungicides were applied at 15 gal/acre and 40 psi using a Lee self-propelled sprayer equipped with a 10-foot boom, fitted with six TJ-VS 8002 nozzles spaced 20 inches apart. Fungicides were applied on July 25 at V14, July 31 at silk (R1), and August 20 at milk (R3) growth stages. Disease ratings were assessed on September 9 at dent (R5) and September 25 at maturity (R6) growth stages. Tar spot stromata severity was visually assessed as a percentage (0–100%) of symptomatic leaf area at ear leaf on five plants per plot and averaged before analysis. Values for the five leaves were averaged before analysis. The two center rows of each plot were harvested on October 18, and yields were adjusted to 15.5% moisture. All disease and yield data were analyzed in SAS 9.4 (SAS Institute, Cary, NC). A generalized linear mixed model analysis of variance was performed using PROC GLIMMIX. Values are least squares means, and values with different letters are significantly different based on Fisher's least significant difference ($\alpha = 0.05$).

In 2024, weather conditions were moderately favorable for the diseases. Tar spot was the most prominent disease and was first detected in plots on August 2. All fungicide programs reduced tar spot on compared to the nontreated control on September 9 and September 25 (Table 27). Fungicide programs of Delaro Complete 12.0 fl oz at V14, Delaro 325 SC 10.0 fl oz at R1, Delaro Complete 8.0 fl oz at R1 followed by Delaro Complete 8.0 fl oz at R3, Delaro Complete 8.0 fl oz at R1 followed by Prosaro Pro 10.3 fl oz at R3, Delaro Complete 8.0 fl oz at R1 followed by Absolute Maxx 6.0 fl oz at R3, and Delaro Complete 8.0 fl oz at R1 followed by Delaro 8.0 fl oz at R3 resulted in the lowest tar spot on September 25. A number of the programs increased harvest moisture over nontreated control (range 17.5% to 21.3%). No significant differences were detected for canopy greenness, test weight, and yield of corn.

TABLE 27. *Effect of Treatment on Tar Spot Severity, Canopy Greenness, and Yield of Corn*

TREATMENT, RATE/ACRE, AND TIMING[z]	TAR SPOT %[y] SEPTEMBER 9	TAR SPOT %[y] SEPTEMBER 25	CANOPY GREEN[x] %	HARVEST MOISTURE %	TEST WEIGHT LB/BU	YIELD[w] BU/ACRE
Nontreated control	1.6 a	22.5 a	51.3	17.5 d	56.2	211.0
Delaro Complete 3.82 SC 8.0 fl oz at V14	0.6 def	6.9 d	51.3	20.1 abc	54.6	209.9
Delaro Complete 3.82 SC 12.0 fl oz at V14	0.4 f	3.0 f	51.3	21.3 a	54.7	207.9
Delaro Complete 3.82 SC 8.0 fl oz at R1	0.7 cd	6.7 de	55.0	19.0 bcd	55.5	216.4
Delaro Complete 3.82 SC 10.0 fl oz at R1	0.6 cde	4.8 def	56.3	19.7 a-d	54.8	218.3
Delaro Complete 3.82 SC 12.0 fl oz at R1	0.5 def	4.1 def	53.8	19.1 bcd	55.2	213.8
Delaro 325 SC 10.0 fl oz at R1	0.5 def	3.3 f	62.5	19.2 a-d	55.9	216.8
Veltyma 7.0 fl oz at R1	0.6 def	3.7 ef	55.0	17.6 d	56.4	217.0
Miravis Neo 13.7 fl oz at R1	0.9 bc	15.4 b	52.5	20.1 abc	55.5	211.4
Trivapro 13.7 fl oz at R1	0.9 bc	11.6 c	66.3	20.0 cd	61.6	214.7
Delaro Complete 3.82 SC 8.0 fl oz at R3	1.1 b	15.3 b	56.3	18.0 ab	56.0	206.7
Delaro Complete 3.82 SC 8.0 fl oz at R1 fb Delaro Complete 3.82 SC 8.0 fl oz at R3	0.4 ef	2.7 f	61.3	20.6 ab	55.3	206.7
Delaro Complete 3.82 SC 8.0 fl oz at R1 fb Prosaro Pro 10.3 fl oz at R3	0.5 def	2.7 f	50.0	18.3 cd	56.3	212.0
Delaro Complete 3.82 SC 8.0 fl oz at R1 fb Absolute Maxx 6.0 fl oz at R3	0.4 ef	2.2 f	57.5	19.0 bcd	55.8	221.2
Delaro Complete 3.82 SC 8.0 fl oz at R1 fb Delaro 325 SC 8.0 fl oz at R3	0.4 ef	2.1 f	50.0	17.6 d	56.3	218.6
P-value[v]	0.0001	0.0001	0.3723	0.0137	0.3980	0.7459

[z] Fungicides were applied on July 25 at V14, July 31 at silk (R1), and August 20 at milk (R3) growth stages and included NIS (Preference) at 0.25% v/v at R1 and R3. fb = followed by.

[y] Tar spot stromata was visually assessed as a percentage (0–100%) of leaf area on five plants in each plot at the ear leaf on September 9 at dent (R5) and September 25 at maturity (R6) growth stages.

[x] Canopy greenness as % visually rated per plot was assessed on September 16 at dent (R5) growth stage.

[w] Yields were adjusted to 15.5% moisture and harvested on October 18.

[v] All data were analyzed in SAS 9.4 (SAS Institute, Cary, NC). A generalized linear mixed model analysis of variance was performed using PROC GLIMMIX. Values are least squares means, and values with different letters are significantly different based on Fisher's least significant difference (α = 0.05).

EVALUATION OF FUNGICIDE FOR TAR SPOT IN CORN IN NORTHWESTERN INDIANA, 2024 (CORN 24-31.PPAC)

S. Shim and D. E. P. Telenko, Department of Botany and Plant Pathology, Purdue University, West Lafayette, IN 47907-2054

CORN (*ZEA MAYS* W2584 VT2P RIB)

Tar spot, *Phyllachora maydis*

A trial was established at the Pinney Purdue Agricultural Center (PPAC) in Porter County, Indiana. The experiment was a randomized complete block design with four replications. Plots were 10 feet wide and 30 feet long and consisted of four rows, and the two center rows were used for evaluation. The previous crop was corn. Standard practices for irrigated grain corn production in Indiana were followed. Corn hybrid W2584 VT2P RIB was planted in 30-inch row spacing at a rate of 34,000 seeds/acre on May 23. The field was overhead irrigated weekly at 1 inch unless weekly rainfall was 1 inch or higher to encourage disease. All products were applied at 15 gal/acre and 40 psi using a Lee self-propelled sprayer equipped with a 10-foot boom, fitted with six TJ-VS 8002 nozzles spaced 20 inches apart. Treatments were applied on August 7 at silk (R1) growth stage. Disease ratings were assessed on September 5 at dent (R5.10) and September 13 at dent (R5.30) growth stages. Tar spot stromata severity was visually assessed as a percentage (0–100%) of symptomatic leaf area at ear leaf (EL) on five plants per plot and averaged before analysis. Values for the five leaves were averaged before analysis. The two center rows of each plot were harvested on October 22, and yields were adjusted to 15.5% moisture. All disease and yield data were analyzed in SAS 9.4 (SAS Institute, Cary, NC). A generalized linear mixed model analysis of variance was performed using PROC GLIMMIX. Values are least squares means, and values with different letters are significantly different based on Fisher's least significant difference (α = 0.05).

In 2024, weather conditions were moderately favorable for the diseases. Tar spot was the most prominent disease in the trial and was first detected in the field on July 19. No significant differences for all treatments were detected for tar spot on September 5 and September 13 at EL (Table 28). The lowest severity of tar spot was with treatments of Veltyma and Veltyma + WC250 4.0 fl oz but were not significantly different from Veltyma + WC460 3.0 fl oz, Veltyma + WC450 3.0 fl oz, and Veltyma + WC112 0.25 % V/V. Veltyma + WC460 3.0 fl oz and Veltyma + WC450 3.0 fl oz significantly increased canopy greenness over nontreated control on September 16. No significant differences were detected for harvest moisture and test weight of corn. All fungicides program increased grain yield over nontreated control except WC634 1.0 GAL/acre + WC450 3.0 fl oz and WC843 1.0 QT/acre + WC450 3.0 fl oz. The treatment of Veltyma + WC250 4.0 fl oz had the highest yield of corn.

TABLE 28. *Effect of Treatment on Tar Spot Severity, Canopy Greenness, and Yield of Corn*

TREATMENT AND RATE/ACRE[z]	TAR SPOT % STROMATA[y] SEPTEMBER 5	TAR SPOT % STROMATA[y] SEPTEMBER 13	CANOPY GREEN[x] %	HARVEST MOISTURE %	TEST WEIGHT LB/BU	YIELD[w] BU/ACRE
Nontreated control	6.7	7.9	68.8 c	16.7	57.7	176.0 c
Veltyma 7.0 fl oz	5.5	6.6	70.0 bc	18.8	57.3	207.2 b
Veltyma 7.0 fl oz + WC250 4.0 fl oz	6.2	6.3	70.0 bc	19.2	56.7	217.6 a
Veltyma 7.0 fl oz + WC460 3.0 fl oz	5.5	4.7	72.5 ab	19.1	56.5	202.4 b
Veltyma 7.0 fl oz + WC450 3.0 fl oz	6.4	7.8	73.8 a	18.5	57.3	200.1 b
Veltyma 7.0 fl oz + WC112 0.25 % V/V	5.0	5.8	71.3 abc	18.9	57.1	197.1 b
WC634 1.0 GAL/acre + WC450 3.0 fl oz	5.5	6.4	70.0 bc	17.6	56.9	173.5 c
WC843 1.0 QT/acre + WC450 3.0 fl oz	4.7	5.7	70.0 bc	17.1	57.3	181.6 c
P-value[v]	0.8485	0.9910	0.0282	0.0691	0.5808	0.0001

[z] Fungicides were applied on August 7 at silk (R1) growth stage.

[y] Tar spot stromata was visually assessed as a percentage (0–100%) of leaf area on five plants in each plot at the ear leaf on September 5 at dent (R5.10) and September 13 at dent (R5.30) growth stages.

[x] Canopy greenness as percent (0–100%), visually rated per plot, was assessed on September 16 at dent (R5) growth stage.

[w] Yields were adjusted to 15.5% moisture and harvested on October 22.

[v] All data were analyzed in SAS 9.4 (SAS Institute, Cary, NC). A generalized linear mixed model analysis of variance was performed using PROC GLIMMIX. Values are least squares means, and values with different letters are significantly different based on Fisher's least significant difference (α = 0.05).

EVALUATING BIOLOGICAL FUNGICIDES FOR TAR SPOT IN CORN IN NORTHWESTERN INDIANA, 2024 (CORN 24-36.PPAC)

E. S. Peña, S. Shim, and D. E. P. Telenko, Department of Botany and Plant Pathology, Purdue University, West Lafayette, IN 47907-2054

CORN (*ZEA MAYS* W2584 VT2P RIB)

Tar spot, *Phyllachora maydis*

A trial was established at the Pinney Purdue Agricultural Center (PPAC) in Porter County, Indiana. The experiment was a randomized complete block design with four replications. Plots were 10 feet wide and 30 feet long and consisted of four rows, and the two center rows were used for evaluation. The previous crop was soybean. Standard practices for nonirrigated grain corn production in Indiana were followed. Corn hybrid W2584 VT2P RIB was planted in 30-inch row spacing at a rate of 34,000 seeds/acre on May 23. Foliar fungicides were applied at 15 gal/acre and 40 psi using a Lee self-propelled sprayer equipped with a 10-foot boom, fitted with six TJ-VS 8002 nozzles spaced 20 inches apart. Fungicides were applied on July 30 at silk (R1) growth stage. Disease ratings were assessed on September 4 and 13 September at dough/dent (R4/R5) and dent (R5) growth stages, respectively. Tar spot stromata severity was visually assessed as a percentage (0–100%) of affected leaf area at ear leaf on five plants per plot and averaged before analysis. The two center rows of each plot were harvested on October 18, and yields were adjusted to 15.5% moisture. All disease and yield data were analyzed in SAS 9.4 (SAS Institute, Cary, NC). A generalized linear mixed model analysis of variance was performed using PROC GLIMMIX. Values are least squares means, and values with different letters are significantly different based on Fisher's least significant difference ($\alpha = 0.05$).

In 2024, weather conditions were moderately favorable for disease. Tar spot was the most prominent disease in the trial and was first detected in plots on July 15. On September 4, all the treatments significantly reduced tar spot severity compared to nontreated control (Table 29). On September 13, no significant differences were detected for tar spot severity. There were no significant differences between treatments for moisture, test weight, and yield of corn.

TABLE 29. *Effect of Treatment on Foliar Disease Severity and Yield of Corn*

TREATMENT AND RATE/ACRE[z]	TAR SPOT % STROMATA[y] SEPTEMBER 4	TAR SPOT % STROMATA[y] SEPTEMBER 13	HARVEST MOISTURE %	TEST WEIGHT LB/BU	YIELD[x] BU/ACRE
Nontreated control	0.7 a	11.9	18.3	56.7	205.5
Headline Amp 10.0 fl oz	0.3 b	4.8	17.5	57.1	212.9
Serifel 16.0 oz	0.4 b	8.7	17.8	60.4	210.0
Actinovate 12.0 oz	0.4 b	6.5	17.6	56.8	209.1
Badge X2 1.8 lb	0.3 b	4.4	17.4	56.4	205.2
Oxidate 2.0 1:100 ratio	0.4 b	7.1	18.3	56.4	206.4
P-value[w]	0.0026	0.1806	0.7820	0.5367	0.8149

[z] Fungicide treatments were applied on July 30 at silk (R1) growth stage.

[y] Tar spot stromata was visually assessed as a percentage (0–100%) of affected leaf area on five plants in each plot at the ear leaf on September 4 and September 13 at dough/dent (R4/R5) and dent (R5) growth stages, respectively.

[x] Yields were adjusted to 15.5% moisture and harvested on October 18.

[w] All data were analyzed in SAS 9.4 (SAS Institute, Cary, NC). A generalized linear mixed model analysis of variance was performed using PROC GLIMMIX. Values are least squares means, and values with different letters are significantly different based on Fisher's least significant difference ($\alpha = 0.05$).

EVALUATION OF FUNGICIDE FOR TAR SPOT IN CORN IN NORTHWESTERN INDIANA, 2024 (CORN 24-37.PPAC)

S. Shim and D. E. P. Telenko, Department of Botany and Plant Pathology, Purdue University, West Lafayette, IN 47907-2054

CORN (*ZEA MAYS* W2584 VT2P RIB)

Tar spot, *Phyllachora maydis*

A trial was established at the Pinney Purdue Agricultural Center (PPAC) in Porter County, Indiana. The experiment was a randomized complete block design with four replications. Plots were 10 feet wide and 30 feet long and consisted of four rows, and the two center rows were used for evaluation. The previous crop was soybean. Standard practices for nonirrigated grain corn production in Indiana were followed. Corn hybrid W2584 VT2P RIB was planted in 30-inch row spacing at a rate of 34,000 seeds/acre on May 23. All fungicides were applied at 15 gal/acre and 40 psi using a Lee self-propelled sprayer equipped with a 10-foot boom, fitted with six TJ-VS 8002 nozzles spaced 20 inches apart. Fungicides were applied on July 30 at silk (R1), August 16 at blister (R2), and August 20 at milk (R3) growth stages. Disease ratings were assessed on August 28 at R3, September 9 at dent (R5), and September 25 at maturity (R6) growth stages. Tar spot stromata severity was visually assessed as a percentage (0–100%) of symptomatic leaf area at ear leaf on five plants per plot and averaged before analysis. Values for the five leaves were averaged before analysis. The two center rows of each plot were harvested on October 18, and yields were adjusted to 15.5% moisture. All disease and yield data were analyzed in SAS 9.4 (SAS Institute, Cary, NC). A generalized linear mixed model analysis of variance was performed using PROC GLIMMIX. Values are least squares means, and values with different letters are significantly different based on Fisher's least significant difference (α = 0.05).

In 2024, weather conditions were moderately favorable for the diseases. Tar spot was the most prominent disease in the trial and was first detected in plots on July 15. Treatments that included Miravis Neo significantly reduced tar spot on August 28 and September 9 over the nontreated control (Table 30). Treatments that included Miravis Neo significantly reduced tar spot on September 25 as compared to nontreated control. CX-9032 1.0 pt at R1 also significantly reduced tar spot on September 25 as compared to the nontreated control. No significant differences were detected for harvest moisture, test weight, and yield of corn.

TABLE 30. *Effect of Treatment on Tar Spot Severity, Lodging, and Yield of Corn*

TREATMENT, RATE/ACRE, AND TIMING[z]	TAR SPOT % STROMATA[y] AUGUST 28	TAR SPOT % STROMATA[y] SEPTEMBER 9	TAR SPOT % STROMATA[y] SEPTEMBER 25	HARVEST MOISTURE %	TEST WEIGHT LB/BU	YIELD[x] BU/ACRE
Nontreated control	0.7 a	2.3 a	24.5 a	18.0	56.4	215.0
CX-9032 1.0 pt at R1	0.7 a	2.1 ab	20.8 bcd	17.5	56.6	227.5
CX-9032 1.0 qt + Miravis Neo 13.7 fl oz at R1	0.2 b	1.8 bc	20.0 cde	18.0	56.7	260.4
CX-10250 1.0 oz at R1	0.6 a	2.0 ab	21.2 a-d	18.2	56.8	224.9
CX-10250 1.0 oz at R1 and R3	0.6 a	2.1 ab	23.8 ab	18.1	56.7	212.4
CX-10250 1.0 oz at R2	0.6 a	2.0 ab	23.3 abc	17.4	56.6	229.8
CX-10250 1.0 oz + Miravis Neo 13.7 fl oz at R1	0.2 b	1.3 d	16.5 e	18.1	56.9	233.1
Miravis Neo 13.7 fl oz at R1	0.1 b	1.4 cd	18.3 de	18.9	61.9	253.1
P-value[w]	0.0001	0.0006	0.0019	0.2949	0.4828	0.3824

[z] Fungicides were applied on July 30 at silk (R1), August 16 at blister (R2), and August 20 at milk (R3) growth stages and included NIS (Preference) at 0.25% v/v.

[y] Tar spot stromata severity was visually assessed as a percentage (0–100%) of leaf area on five plants in each plot at the ear leaf on August 28 at milk (R3), September 9 at dent (R5) and September 25 at maturity (R6) growth stages, respectively.

[x] Yields were adjusted to 15.5% moisture and harvested on October 18.

[w] All data were analyzed in SAS 9.4 (SAS Institute, Cary, NC). A generalized linear mixed model analysis of variance was performed using PROC GLIMMIX. Values are least squares means, and values with different letters are significantly different based on Fisher's least significant difference (α = 0.05).

FUNGICIDE COMPARISON FOR TAR SPOT ON SHORT CORN IN NORTHWESTERN INDIANA, 2024 (COR24-39.PPAC)

E. Schillinger, S. Shim, and D. E. P. Telenko, Department of Botany and Plant Pathology, Purdue University, West Lafayette, IN 47907-2054

CORN (*ZEA MAYS* PR111-20SSC AND PR108-20SSC)

Tar spot, *Phyllachora maydis*

A trial was established at the Pinney Purdue Agricultural Center (PPAC) in Porter County, Indiana. The experiment was a randomized complete block design with four replications. Plots were 10 feet wide and 30 feet long and consisted of four rows, and the two center rows were used for evaluation. The previous crop was corn. Standard practices for irrigated grain corn production in Indiana were followed. Corn hybrid PR111-20SSC and PR108-20SSC were planted in 30-inch row spacing at a rate of 2 seeds/foot on May 22. All fungicides were applied at 15 gal/acre and 40 psi using a Lee self-propelled sprayer equipped with a 10-foot boom, fitted with six TJ-VS 8002 nozzles spaced 20 inches apart. Fungicides were applied on August 7 at silk (R1) growth stage. Disease ratings were assessed on September 4 and September 13 at dough/dent (R4/R5) and dent (R5) growth stages, respectively. Tar spot stromata severity and chlorosis/necrosis were visually assessed as a percentage (0–100%) of symptomatic leaf area at ear leaf on five plants per plot and averaged before analysis. Values for the five leaves were averaged before analysis. The two center rows of each plot were harvested on October 23, and yields were adjusted to 15.5% moisture. All disease and yield data were analyzed in SAS 9.4 (SAS Institute, Cary, NC). A generalized linear mixed model analysis of variance was performed using PROC GLIMMIX. Values are least squares means, and values with different letters are significantly different based on Fisher's least significant difference (α = 0.05).

In 2024, weather conditions were moderately favorable for the diseases. Tar spot was the most prominent disease in the trial and was first detected in plots on July 19. There was no significant difference between hybrids on tar spot severity on September 4 and 13 (Table 31). On September 13, Hybrid PR111-20SSC significantly reduced tar spot chlorosis and necrosis symptoms compared to PR108-20SSC. Harvest moisture was significantly higher for PR111-20SSC than PR108-20SSC. There was no significant difference between hybrids for canopy greenness, test weight, and yield. The Delaro Complete application significantly reduced tar spot severity over the nontreated control on September 4 and 13 and significantly reduced chlorosis and necrosis compared to September 13. Delaro Complete significantly increased harvest moisture over the nontreated control; there was no significant difference in test weight. Delaro Complete application significantly increased the yield of corn over the nontreated control.

TABLE 31. *Effect of Treatment on Tar Spot Severity, Canopy Greenness, and Yield of Corn*

HYBRID, TREATMENT, AND RATE/ACRE[z]	TAR SPOT % STROMATA[y] SEPTEMBER 4	TAR SPOT % STROMATA[y] SEPTEMBER 13	TAR SPOT % CHLOR/NEC[x] SEPTEMBER 13	CANOPY GREEN[w] %	HARVEST MOISTURE %	TEST WEIGHT LB/BU	YIELD[v] BU/ACRE
Hybrid							
PR111-20SSC	0.6	5.6	0.4 b	76.9	21.3 a	54.0	176.1
PR108-20SSC	0.7	5.3	2.3 a	73.1	18.9 b	54.5	171.3
Fungicide							
Nontreated control	1.2 a	9.5 a	2.6 a	70.6	18.7 b	54.5	162.9 b
Delaro Complete 3.82 SC 8.0 fl oz	0.2 b	1.5 b	0.1 b	79.4	21.5 a	54.0	184.5 a
P-value hybrid[u]	0.4312	0.4695	0.0009	0.4250	0.0314	0.4458	0.4941
P-value fungicide	0.0001	0.0001	0.0001	0.0830	0.0155	0.3816	0.0105
P-value hybrid*fungicide	0.8727	0.3917	0.0022	0.1290	0.3812	0.1948	0.7017

[z] Delaro Complete was applied on August 7 at silk (R1) growth stage.

[y] Tar spot stromata was visually assessed as a percentage (0–100%) of leaf area on five plants in each plot at the ear leaf on September 4 and September 13 at dough/dent (R4/R5) and at dent (R5) growth stages, respectively.

[x] Tar spot chlorosis and necrosis was visually assessed as a percentage (0–100%) of leaf area on five plants in each plot at the ear leaf on September 13 at dent (R5) growth stage.

[w] Canopy greenness as %, visually rated per plot was assessed on September 16 at dent (R5) growth stage.

[v] Yields were adjusted to 15.5% moisture and harvested on October 23.

[u] All data were analyzed in SAS 9.4 (SAS Institute, Cary, NC). A generalized linear mixed model analysis of variance was performed using PROC GLIMMIX. Values are least squares means, and values with different letters are significantly different based on Fisher's least significant difference (α = 0.05).

EVALUATION OF DENT (R5) FUNGICIDE APPLICATION IN CORN IN NORTHWESTERN INDIANA, 2024 (COR24-40.PPAC)

S. Shim and D. E. P. Telenko, Department of Botany and Plant Pathology, Purdue University, West Lafayette, IN 47907-2054

CORN (*ZEA MAYS* W2584VT2PRIB)

Tar spot, *Phyllachora maydis*

A trial was established at the Pinney Purdue Agricultural Center (PPAC) in Porter County, Indiana. The experiment was a randomized complete block with eight replications. Plots were 10 feet wide and 30 feet long and consisted of four rows, and the two center rows were used for evaluation. The previous crops were soybean. Standard practices for nonirrigated grain corn production in Indiana were followed. Corn hybrid W2584VT2PRIB was planted in 30-inch row spacing at a rate of 34,000 seeds/foot on May 23. Fungicide application was done at 15 gal/acre and 40 psi using a Lee self-propelled sprayer equipped with a 10-foot boom, fitted with six TJ-VS 8002 nozzles spaced 20 inches apart. Fungicide was applied on September 12 at t dent (R5) growth stage. Disease ratings were assessed on September 25 at maturity (R6) growth stage. Tar spot severity was rated by visually assessing the percentage of stroma per leaf on 10 plants in each plot at the ear leaf. Values for each plot were averaged before analysis. Lodging was evaluated on September 25 by determining the percentage of lodged stalks when pushed from shoulder height to 45° from vertical. The two center rows of each plot were harvested on October 18, and yields were adjusted to 15.5% moisture. All data were analyzed in SAS 9.4 (SAS Institute, Cary, NC). A generalized linear mixed model analysis of variance was performed using PROC GLIMMIX. Values are least squares means, and values with different letters are significantly different based on Fisher's least significant difference ($\alpha = 0.05$).

In 2024, weather conditions were favorable for disease. Tar spot was the most prominent in the trial and reached moderate severity in the trial. There was no significant difference in treatments at R5 for tar spot severity (Table 32). There was no significant effect on treatments on lodging, harvest moisture, test weight, and yield of corn.

TABLE 32. *Effect of Fungicide Application at Dent (R5) for Tar Spot Severity, Lodging, and Yield of Corn*

TREATMENT, RATE/ACRE, AND TIMING[z]	TAR SPOT[y] %	LODGING[x] %	HARVEST MOISTURE %	TEST WEIGHT LB/BU	YIELD[w] BU/ACRE
Nontreated control	23.0	15.0	16.9	57.1	219.1
Veltyma 3.34 SC 7.0 fl oz at R5	23.2	25.0	16.7	57.1	209.5
P-value[v]	0.8715	0.2753	0.6551	0.9683	0.1832

[z] Veltyma application was made at 15 gal/acre and 40 psi using a Lee self-propelled sprayer equipped with a 10-foot boom, fitted with six TJ-VS 8002 nozzles spaced 20 inches apart. Fungicide was applied on September 12 at dent (R5) growth stage and contained a nonionic surfactant at a rate of 0.25% v/v.

[y] Tar spot stroma was visually assessed as a percentage (0–100%) of leaf area on 10 plants in each plot at the ear leaf on September 25 at R6 (maturity) growth stage.

[x] Lodging was determined as percentage of lodged stalks when pushed from shoulder height to 45° from vertical on September 25.

[w] Yields were adjusted to 15.5% moisture and harvested on October 18.

[v] All data were analyzed in SAS 9.4 (SAS Institute, Cary, NC). A generalized linear mixed model analysis of variance was performed using PROC GLIMMIX. Values are least squares means, and values with different letters are significantly different based on Fisher's least significant difference (α = 0.05).

EVALUATION OF SEED TREATMENT FOR SUDDEN DEATH SYNDROME ON SOYBEAN IN NORTHWESTERM INDIANA, 2024 (SOY24-04.PPAC)

E. Myers, S. Shim, and D. E. P. Telenko, Department of Botany and Plant Pathology, Purdue University, West Lafayette, IN 47907-2054

SOYBEAN (*GLYCINE MAX* AG26XF1)

Sudden death syndrome, *Fusarium virguliforme*

A trial was established at the Purdue Pinney Agricultural Center (PPAC) in Porter County, Indiana. The experiment was a randomized complete block design with four replications. Plots were 10 feet wide and 30 feet long and consisted of four rows, and the two center rows were used for evaluation. The previous crop was corn. Soybean cultivar AG26XF1 was planted in 30-inch row spacing at a rate of 8 seeds/foot on April 25. *Fusarium virguliforme* inoculum was applied at planting at 1.25 g/foot within the seedbed. Seed treatments were applied on seeds before planting. Standard practices for soybean production in Indiana were followed. In-furrow and 2x2 applications were applied in 10 gal/acre at planting on April 25. Xylem Plus was applied at full flowering (R2) growth stage at 15 gal/acre and 40 psi using a Lee self-propelled sprayer equipped with a 10-foot boom, fitted with six TJ-VS 8002 nozzles spaced 20 inches apart. Disease ratings were assessed on August 30 at beginning maturity (R7) growth stage. SDS in each plot was rated for disease incidence (DI) and disease severity (DS). Disease incidence was percentage of plants with disease symptoms, and disease severity (DS) was rated using a scale of 1–9, where 1 refers to low disease pressure and 9 refers to premature death of the plant. The SDS index was then calculated using the equation $DX = (DI \times DS)/9$. Root rot rating was assessed at R4 (full pod) growth stage on August 9. The two center rows of each plot were harvested on October 2, and yields were adjusted to 13% moisture. All disease and yield data were analyzed in SAS 9.4 (SAS Institute, Cary, NC). A generalized linear mixed model analysis of variance was performed using PROC GLIMMIX. Values are least squares means, and values with different letters are significantly different based on Fisher's least significant difference ($\alpha = 0.05$).

In 2024, weather conditions were moderately favorable for disease development. SDS was present in the trial. All the treatments significantly reduced SDS incidence and index compared to the nontreated control (Table 33). There was no significant effect of treatments on root rot severity and test weight compared to the nontreated control. All the seed treatments significantly increased the yield compared to the nontreated control.

TABLE 33. *Effect of Seed Treatment on SDS, Root Rot, and Yield of Soybean*

TREATMENT, RATE/ACRE, AND TIMING[z]	SDS DI[y]	SDS INDEX[x]	ROOT ROT %[w]	TEST WEIGHT LB/BU	YIELD[v] BU/ACRE
Susceptible + nontreated control	42.5 a	10.8 a	6.6	56.5	46.2 c
Susceptible + base	13.8 b	2.6 b	6.8	56.5	56.0 a
Susceptible + IleVO	12.5 b	2.2 b	3.8	56.8	53.2 ab
Susceptible + Saltro	12.5 b	1.4 b	2.5	56.5	54.8 ab
Susceptible + Zeltera	17.5 b	3.1 b	7.0	56.5	53.5 ab
Susceptible, base fb Xylem Plus in-furrow 32.0 fl oz/acre fb Xylem Plus 24.0 fl oz at R2	17.5 b	3.3 b	5.4	56.7	53.9 ab
Susceptible, base fb Xyway 2x2 15.2 fl oz	13.8 b	2.1 b	6.2	56.8	50.6 b
Susceptible + base + ILeVO + Ceramax	18.8 b	2.8 b	5.6	56.7	54.3 ab
P-value[u]	0.0031	0.0004	0.3002	0.4563	0.0030

[z] Seed treatments were applied on seeds before planting. In-furrow and 2x2 application were applied in 10 gal/acre at planting on April 25. Xylem Plus was applied at full flowering (R2) growth stage. fb = followed by.

[y] SDS in each plot was rated for disease incidence (DI) as a percentage of plants with disease symptoms (0–100%) on August 30 at R7 (beginning maturity) growth stage.

[x] SDS index (DX) calculated using the equation DX = (DI*DS/9). SDS = sudden death syndrome.

[w] Root rot rating was assessed at R4 (full pod) growth stage on August 9. Ten roots per plot were sampled from border rows, gently washed, and rated for root rot severity on a scale of 0–100%.

[v] Yields were adjusted to 13% and harvested on October 2.

[u] All data were analyzed in SAS 9.4 (SAS Institute, Cary, NC). A generalized linear mixed model analysis of variance was performed using PROC GLIMMIX. Values are least squares means, and values with different letters are significantly different based on Fisher's least significant difference (α = 0.05).

COMPARISON OF PLANTING DATES AND SEED TREATMENTS ON SOYBEAN IN NORTHWESTERN INDIANA, 2024 (SOY24-09.PPAC)

I. L. Miranda, S. Shim, and D. E. P. Telenko, Department of Botany and Plant Pathology, Purdue University, West Lafayette, IN 47907-2054

SOYBEAN (*GLYCINE MAX* 24E453N)

Septoria brown spot, *Septoria glycines*

A trial was established at Pinney Purdue Agricultural Center (PPAC) in Wanatah, Indiana. The experiment design was a split plot with four replications. The main plot was planting date and subplot seed treatments. Plots were 10 feet wide and 30 feet long and consisted of four rows, and the two center rows were used for evaluation. The previous crop was corn. Standard practices for soybean production in Indiana were followed. Soybean cultivar 24E453N was planted in 30-inch row spacing at a rate of 8 seeds/foot. Treatments were a factorial arrangement of four planting dates for seed treatments. Soybeans were planted on April 15 (planting date 1), May 2 (planting date 2), May 23 (planting date 3) and May 31 (planting date 4). Stand counts were assessed at cotyledons expanded/first-node (VC/V1) growth stage for each planting date. Disease ratings were assessed on August 31 at full seed/beginning maturity/full maturity (R6/R7/R8) growth stage. Septoria brown spot (SBS) was rated for disease severity by visually assessing the percentage of symptomatic leaf area in the lower canopy. Ten roots were sampled for outer rows of each plot and rated for root rot severity on a scale of 0–100% and averaged before analysis. The two center rows of each plot were harvested on September 30, and yields were adjusted to 13% moisture. All data were analyzed in SAS 9.4 (SAS Institute, Cary, NC). A generalized linear mixed model analysis of variance was performed using PROC GLIMMIX. Values are least squares means, and values with different letters are significantly different based on Fisher's least significant difference ($\alpha = 0.05$).

In 2024, weather conditions were not favorable for the disease, and little disease developed in plots. Main effects of planting date and seed treatment are presented, since there was only an interaction for stand count and not the other dependent variables. Soybean stand was the highest at planting on May 23 compared to the other planting dates (Table 34). SBS was lowest at the latest planting date on May 31. The planting dates of April 15 and May 2 significantly increased root rot over planting at May 23 and May 31. Harvest moisture was highest at planting on May 2 but was not significantly different from the planting date of April 15. Planting on May 23 and May 31 increased test weight over the earlier planting, while the highest soybean yield occurred when planted on April 15 or May 2. CruiserMaxx APX with and without Thiamethoxam increased soybean stand compared to the nontreated control and Thiamethoxam-only seed treatment. Seed treatment of CruiserMaxx + Thiamethoxan had significantly higher SBS severity compared to the nontreated control and Thiamethoxam. No significant differences were observed between seed treatments for root rot and harvest moisture. CruiserMaxx APX without Thiamethoxam and the nontreated control had a higher test weight compared to CruiserMaxx APX + Thiamethoxam. No significant interactions between seed treatments were observed on yield of soybean.

TABLE 34. *Effect of Planting Dates and Seed Treatments on Stand Count, SBS, Root Rot, and Yield of Soybean*

PLANTING DATES AND SEED TREATMENTS[z]	STAND COUNT #/ACRE	SBS[y] %	ROOT ROT[x] %	MOISTURE %	TEST WEIGHT LB/BU	YIELD[w] BU/ACRE
Planting date						
Planting date 1 (15 Apr)	89,897 c	5.9 a	7.7 a	14.2 ab	55.8 c	56.9 a
Planting date 2 (2 May)	103,673 b	6.6 a	5.1 b	14.4 a	56.3 b	54.9 a
Planting date 3 (23 May)	133,893 a	6.8 a	3.1 c	13.9 c	56.9 a	48.0 b
Planting date 4 (31 May)	110,751 b	2.6 b	2.4 c	14.1 bc	56.7 a	45.7 b
Seed treatment						
Nontreated control	106,450 b	5.1 b	4.7	14.1	56.5 a	51.8
CruiserMaxx APX + Thiamethoxam	118,864 a	5.6 ab	3.8	14.2	56.2 b	52.0
Thiamethoxam	97,357 c	4.8 b	5.2	14.2	56.4 ab	51.1
CruiserMaxx APX without Thiamethoxam	115,543 a	6.6 a	4.7	14.1	56.6 a	50.5
P-value planting date[v]	*0.0001*	*0.0001*	*0.0001*	*0.0025*	*0.0001*	*0.0001*
P-value seed treatment	*0.0001*	*0.0083*	*0.3086*	*0.7814*	*0.0185*	*0.5842*
P-value planting date*seed treatment	*0.0059*	*0.1634*	*0.9500*	*0.6374*	*0.1250*	*0.7377*

[z] Seed treatments applied prior to planting at 10 g AI/100 kg seed.

[y] Foliar disease severity was rated on a scale of 0–100% of symptomatic leaf area in the lower canopy. SBS = Septoria brown spot.

[x] Root rot was visually assessed as a percentage (0–100%) of dark discoloration on 10 roots per plot.

[w] Yields were adjusted to 13% moisture and harvested on October 3.

[v] All data were analyzed in SAS 9.4 (SAS Institute, Cary, NC). A generalized linear mixed model analysis of variance was performed using PROC GLIMMIX. Values are least squares means, and values with different letters are significantly different based on Fisher's least significant difference (α = 0.05).

FUNGICIDE EVALUATION FOR WHITE MOLD IN SOYBEAN IN NORTHWESTERN INDIANA, 2024 (SOY24-14.PPAC)

C. Rocco da Silva, S. Shim, and D. E. P. Telenko, Department of Botany and Plant Pathology, Purdue University, West Lafayette, IN 47907-2054

SOYBEAN (*GLYCINE MAX* P29A19E)

White mold, *Sclerotinia sclerotiorum*

A trial was established at the Purdue Agricultural Center (PPAC) in Porter County, Indiana. The experiment was a randomized complete block design with four replications. Plots were 10 feet wide and 30 feet long and consisted of four rows, and the two center rows were used for evaluation. The previous crop was corn. Standard practices for soybean production in Indiana were followed. Soybean variety P29A19E was planted in 30-inch row spacing at a rate of 8 seeds/foot on May 22. Fungicides were applied at 15 gal/acre and 40 psi using a CO_2 backpack sprayer at beginning bloom (R1) and a Lee self-propelled sprayer equipped with a 10-foot boom, fitted with six TJ-VS 8002 nozzles spaced 20 inches apart, at 5 mph at R2 and R3. Fungicides were applied on July 12 at R1, July 19 at full bloom (R2), and July 25 at beginning pod (R3) growth stages. The two center rows of each plot were harvested on October 2, and yields were adjusted to 13% moisture. All data were analyzed in SAS 9.4 (SAS Institute, Cary, NC). A generalized linear mixed model analysis of variance was performed using PROC GLIMMIX. Values are least squares means, and values with different letters are significantly different based on Fisher's least significant difference ($\alpha = 0.05$).

In 2024, weather conditions were not favorable for disease development. No white mold developed in plots. There was no significant effect of treatment on percent canopy greenness, defoliation, harvest moisture, test weight, and yield of soybean (Table 35).

TABLE 35. *Effect of Treatment on Canopy Greenness, Defoliation, and Yield of Soybean*

TREATMENT, RATE/ACRE, AND TIMING[z]	CANOPY GREEN[y] %	DEFOLIATION[x] %	HARVEST MOISTURE %	TEST WEIGHT LB/BU	YIELD[w] BU/ACRE
Nontreated control	16.3	13.8	13.9	55.4	62.1
Endura 70 WDG 8.0 oz at R1	11.8	18.8	13.9	55.4	60.1
Endura 70 WDG 8.0 oz at R2	18.8	12.5	13.9	55.2	57.2
Endura 70 WDG 8.0 oz at R3	24.3	10.0	13.9	55.1	58.4
Endura 70 WDG 8.0 oz at Sporecaster (R2)	20.5	13.8	13.8	55.3	62.3
P-value[v]	*0.2405*	*0.1465*	*0.8776*	*0.9529*	*0.6223*

[z] Fungicides were applied on July 12 at beginning bloom (R1), July 19 at full bloom (R2), and July 25 at beginning pod (R3) growth stages.

[y] Canopy greenness was visually assessed as a percentage (0–100%) of crop canopy green on September 16.

[x] Defoliation was rated on a scale of 0–100% in plot on September 16.

[w] Yields were adjusted to 13% moisture and harvested on October 2.

[v] All data were analyzed in SAS 9.4 (SAS Institute, Cary, NC). A generalized linear mixed model analysis of variance was performed using PROC GLIMMIX. Values are least squares means, and values with different letters are significantly different based on Fisher's least significant difference (α = 0.05).

EVALUATION OF SEED TREATMENTS FOR SUDDEN DEATH SYNDROME IN NORTHWESTERN INDIANA (SOY24-15.PPAC)

J. D. Peña, S. Shim, and D. E. P. Telenko, Department of Botany and Plant Pathology, Purdue University, West Lafayette, IN 47907-2054

SOYBEAN (*GLYCINE MAX* AG26XF1)

Sudden death syndrome, *Fusarium virguliforme*

A trial was established at the Pinney Purdue Agricultural Center (PPAC) in Porter County, Indiana. The experiment was a randomized complete block design with four replications. Plots were 10 feet wide and 30 feet long and consisted of four rows, and the two center rows were used for evaluation. The previous crop was corn. Standard practices for soybean production in Indiana were followed. Soybean cultivar AG26XF1 was planted in 30-inch row spacing at a rate of 8 seeds/foot on April 25. Seed treatment was applied by the cooperator. Inoculum of *Fusarium virguliforme* was applied in the seedbed at 1.25 g/foot at planting. Stand count was assessed on May 28 at V2/V3 growth stage. Ten roots were sampled from outer rows of each plot on August 9 at full pod (R4) growth stage, gently washed, and dried to determine root dry weight (g). The two center rows of each plot were harvested on October 2, and yields were adjusted to 13% moisture. A generalized linear mixed model analysis of variance was performed using PROC GLIMMIX. Values are least squares means, and values with different letters are significantly different based on Fisher's least significant difference (α = 0.05).

In 2024, weather conditions were not favorable for the disease; only low levels of sudden death syndrome (SDS) were detected. Seed treatments of Base + Ilevo and Base alone increased soybean stand at V2/V3 (Table 36). There were no significant differences between treatments for SDS incidence and severity (data not shown). All seed treatments increased root dry weight over nontreated control except Base alone, Base + CeraMax + Germate Plus, and Base + TBZ + Headsup + Biost 2nd Gen + Ascribe SAR. No significant differences were detected between treatments for green stem, moisture, test weight, and yield of soybean.

TABLE 36. *Effect of Seed Treatment on Stand Count, Root Dry Weight, Green Stem, and Yield of Soybean*

TREATMENT[z]	STAND COUNT[y] #/ACRE	ROOT DRY WEIGHT[x] G	GREEN STEM[w] %	HARVEST MOISTURE %	TEST WEIGHT LB/BU	YIELD[v] BU/ACRE
Nontreated control	91,726 cde	16.7 c	1.0	15.3	55.4	47.3
Base	105,452 ab	19.5 bc	0.8	15.3	55.1	51.0
Base + CeraMax + Germate Plus	97,608 bc	21.7 abc	1.0	15.3	55.5	44.4
Base + Avodigen + Adaplan + Ethos Elite	84,754 e	27.6 a	1.0	15.3	55.7	47.4
Base + TBZ + Headsup + Biost 2nd Gen + Ascribe SAR	87,586 de	18.9 bc	0.5	15.3	55.6	45.2
Base + Avodigen + Adaplan + Ethos Elite + TBZ + Headsup + BioST + Ascribe + CeraMax + Germate Plus	94,776 cd	24.7 ab	1.0	15.2	55.3	48.2
Base + ILEVO	111,552 a	23.1 ab	1.8	15.3	55.4	53.0
P-value[u]	0.0001	0.0200	0.2778	0.9352	0.6372	0.1830

[z] Seed treatments were applied prior to planting by the cooperator. All plots were inoculated with *Fusarium virguliforme* at planting on April 25.

[y] Stand count was assessed on May 28 at V2/V3 growth stage.

[x] Green stem was rated as a percent of plot (0–100%) just prior to harvest on October 2.

[w] Ten roots were sampled from the outer rows of each plot on August 9 at full pod (R4) growth stage, gently washed, and dried to determine root dry weight (g).

[v] Yields were adjusted to 13% moisture and harvested on October 2.

[u] All data were analyzed in SAS 9.4 (SAS Institute, Cary, NC). A generalized linear mixed model analysis of variance was performed using PROC GLIMMIX. Values are least squares means, and values with different letters are significantly different based on Fishers least significant difference (α = 0.05).

EVALUATION OF PLANTING DATE AND SULFUR FOR SUDDEN DEATH SYNDROME ON SOYBEAN IN NORTHWESTERN INDIANA, 2024 (SOY24-20.PPAC)

E. A. Duncan, S. Shim, S. Casteel, and D. E. P. Telenko, Department of Botany and Plant Pathology, Purdue University, West Lafayette, IN 47907-2054

SOYBEAN (*GLYCINE MAX*)

Sudden death syndrome, *Fusarium virguliforme*

A trial was established at the Pinney Purdue Agricultural Center (PPAC) in Porter County, Indiana. The experimental design was split plot with four replications. The main plot was planting date (April and May), and subplots were a factorials arrangement of inoculation (non-inoculated and inoculated) by treatment (non-treated control, ammonium sulfate, ammonium thiosulfate, and calcium sulfate). Plots were 10 feet wide and 30 feet long and consisted of four rows, and the two center rows were used for evaluation. The previous crop was corn. Standard practices for soybean production in Indiana were followed. Soybean seeds were planted in 30-inch row spacing at a rate of 8 seeds/foot. Soybeans were planted on April 25 (April planting) and on May 22 (May planting). *Fusarium virguliforme* was inoculated at planting at 1.25 g/foot. Sulfur treatments were applied on April 26 and May 23 following planting, with a resultant sulfur rate of 20 lb/acre. Ammonium sulfate at 83 lb/acre and calcium sulfate at 117 lb/acre were hand-applied. Ammonium thiosulfate was applied at 6.9 gal/acre and at 15 gal/acre at 28–29 psi using a CO_2 backpack sprayer equipped with a 10-foot boom, fitted with eight TJ-VS 8002 nozzles spaced 15 inches apart, at 3 mph. Disease ratings were assessed on September 4 at full seed (R6) growth stage. Sudden death syndrome (SDS) in each plot was rated for disease incidence (DI) a percentage of plants with disease symptoms (0–100%) and disease severity (DS) on a scale of 1–9, where, 1 refers to low disease pressure and 9 refers to premature death of the plant. A SDS index (DX) was then calculated using the equation DX = (DI x DS/9). The two center rows of each plot were harvested on October 2, and yields were adjusted to 13% moisture. All data were analyzed in SAS 9.4 (SAS Institute, Cary, NC). A generalized linear mixed model analysis of variance was performed using PROC GLIMMIX. Values are least squares means, and values with different letters are significantly different based on Fisher's least significant difference ($\alpha = 0.05$).

In 2024, weather conditions were not favorable for the disease. SDS was present in the trial and reached low severity. There were significant interactions between inoculum and treatment but no significant interactions with planting date except for SDS severity (DS); therefore, main effects of planting date and subeffects of inoculum by treatment were assessed (Table 37). SDS incidence (DI) and SDS index were significantly lower in soybeans planted on May 22 compared to April 25. Planting on May 22 significantly increased canopy greenness on September 4. There was no significant difference between planting dates for harvest moisture and soybean yield. There were no significant differences between inoculation and treatments for SDS severity, SDS index, canopy greenness, harvest moisture, and yield of soybean.

TABLE 37. *Effect of Planting Date, Inoculation, Sulfur Treatment on and SDS, Canopy Greenness, and Yield of Soybean*

TREATMENTS AND RATE[z]	SDS DI[y]	SDS INDEX[x]	CANOPY GREEN %	HARVEST MOISTURE %	YIELD[w] BU/ACRE
Planting Date					
April planting (April 25)	3.5 a	0.6 a	75.3 b	13.3	75.7
May planting (May 22)	0.9 b	0.1 b	91.7 a	13.3	76.2
Treatment					
Non-inoculated; Nontreated control	2.8	0.5	83.0	13.0	75.8
Non-inoculated; Ammonium sulfate 83 lb	1.3	0.2	85.3	13.4	77.0
Non-inoculated; Ammonium thiosulfate 6.9 gal	1.6	0.2	80.7	13.3	73.5
Non-inoculated; Calcium sulfate 117 lb	1.8	0.2	83.8	13.6	74.4
Inoculated; Nontreated control	3.3	0.6	82.2	13.3	75.1
Inoculated; Ammonium sulfate 83 lb	2.0	0.3	84.1	13.4	76.3
Inoculated; Ammonium thiosulfate 6.9 gal	2.5	0.4	85.2	13.1	77.6
Inoculated; Calcium sulfate 117 lb	2.3	0.4	83.7	13.2	77.4
P-value planting date[v]	*0.0003*	*0.0005*	*0.0016*	*0.8608*	*0.7022*
P-value inoculum	*0.0505*	*0.1009*	*0.5865*	*0.8103*	*0.3495*
P-value sulfur treatment	*0.0314*	*0.0108*	*0.5604*	*0.8093*	*0.9291*
*P- value planting date*treatment*	*0.7795*	*0.2759*	*0.1286*	*0.7034*	*0.7754*
*P-value planting date*inoculum*	*0.3868*	*0.1928*	*0.8107*	*0.8182*	*0.2544*
*P-value inoculum*treatment*	*0.9667*	*0.9160*	*0.2531*	*0.7089*	*0.5205*
*P-value planting date*inoculum*treatment*	*0.7396*	*0.7354*	*0.0645*	*0.6692*	*0.7659*

[z] *Fusarium virguliforme* grown on sorghum was inoculated at planting. Sulfur treatments were applied by hand following planting with a resultant sulfur rate of 20 lb/acre.

[y] SDS in each plot was rated for disease incidence (DI) as a percentage of plants with disease symptoms (0–100%).

[x] SDS index (DX) was calculated using the equation DX = (DI x DS/9).

[w] Yields were adjusted to 13% moisture and harvested on October 2.

[v] All data were analyzed in SAS 9.4 (SAS Institute, Cary, NC). A generalized linear mixed model analysis of variance was performed using PROC GLIMMIX. Values are least squares means, and values with different letters are significantly different based on Fisher's least significant difference (α = 0.05).

FUNGICIDE EVALUATION FOR WHITE MOLD IN SOYBEAN IN NORTHWESTERN INDIANA, 2024 (SOY24-25.PPAC)

C. Rocco da Silva, S. Shim, and D. E. P. Telenko, Department of Botany and Plant Pathology, Purdue University, West Lafayette, IN 47907-2054

SOYBEAN (*GLYCINE MAX* P29A19E)

White mold, *Sclerotinia sclerotiorum*

A trial was established at the Pinney Purdue Agricultural Center (PPAC) in Porter County, Indiana. The experiment was a randomized complete block design with four replications. Plots were 10 feet wide and 30 feet long and consisted of four rows, and the two center rows were used for evaluation. The previous crop was corn. Standard practices for soybean production in Indiana were followed. Soybean cultivar P29A19E was planted in 30-inch row spacing at a rate of 8 seeds/foot on May 22. Fungicides were applied on July 12 at the beginning bloom (R1) and July 25 at the beginning pod (R3) growth stages. Fungicides were applied at 15 gal/acre and 40 psi using a CO_2 backpack sprayer at R1 and a Lee self-propelled sprayer equipped with a 10-foot boom, fitted with six TJ-VS 8002 nozzles spaced 20 inches apart, at 5 mph at R3. Disease ratings were rated on September 4. White mold disease incidence was assessed by counting the number of plants out of 30 in each plot with symptoms. For disease severity, each plant that was observed was rated according to the following disease category: 0 = no disease, 1 = lateral branches with white mycelium and lesions, 2 = main stem with white mycelium and sclerotia present, and 3 = entire plant wilted/plant death. White mold disease severity index: DSI = Total of severity score (0–3) from 30 plants/0.9. The two center rows of each plot were harvested on October 2, and yields were adjusted to 13% moisture. A generalized linear mixed model analysis of variance was performed using PROC GLIMMIX. Values are least squares means, and values with different letters are significantly different based on Fisher's least significant difference ($\alpha = 0.05$).

In 2024, weather conditions were not favorable for disease. White mold was the most prominent disease and reached low severity. No significant differences were detected between fungicide programs and nontreated control for white mold DSI except for Miravis Neo at R1 (Table 38). There was no significant effect of treatment on the canopy greenness, harvest moisture, test weight, and yield of soybean.

TABLE 38. *Effect of Treatment on White Mold, Canopy Greenness, and Yield of Soybean*

TREATMENT, RATE/ACRE, AND TIMING[z]	WHITE MOLD DSI[y] %	CANOPY GREEN[x] %	HARVEST MOISTURE %	TEST WEIGHT LR/BU	YIELD[w] BU/ACRE
Nontreated control	0.0 b	7.8	13.9	55.7	58.0
Delaro Complete 3.82 SC 8.0 fl oz at R1	0.3 b	10.0	14.0	55.4	61.0
Delaro Complete 3.82 SC 8.0 fl oz at R1 fb Delaro Complete 3.82 SC 8.0 fl oz at R3	0.0 b	7.8	14.0	55.1	61.3
Propulse 3.34 SC 8.0 fl oz at R1	0.0 b	6.5	13.8	55.5	59.6
Propulse 3.34 SC 8.0 fl oz at R1 fb Delaro Complete 3.82 SC 8.0 fl oz at R3	0.0 b	13.8	13.9	55.6	59.0
Viatude SC 10.0 fl oz at R1	0.0 b	21.3	13.9	55.4	57.3
Miravis Neo 2.5 SC 13.7 fl oz at R1	1.4 a	6.3	14.0	55.3	59.6
Endura 0.7 DF 6.0 fl oz at R1	0.3 b	17.5	13.9	55.6	59.3
P-value[v]	*0.0333*	*0.1121*	*0.2534*	*0.4722*	*0.8076*

[z] Fungicides were applied on July 12 at the R1 (beginning bloom) and July 25 at the R3 (beginning pod) growth stages, and all treatments contained a nonionic surfactant (Preference) at a rate of 0.25% v/v. fb = followed by.

[y] White mold disease severity index: DSI = Total of severity score (0–3) from 30 plants/0.9.

[x] Canopy greenness was visually assessed as a percentage (0–100%) of crop canopy green on September 16.

[w] Yields were adjusted to 13% moisture and harvested on October 2.

[v] All data were analyzed in SAS 9.4 (SAS Institute, Cary, NC). A generalized linear mixed model analysis of variance was performed using PROC GLIMMIX. Values are least squares means, and values with different letters are significantly different based on Fisher's least significant difference (α = 0.05).

EVALUATING BIOLOGICALS FOR DISEASE MANAGEMENT SOYBEAN IN NORTHWESTERN INDIANA, 2024 (SOY24-26.PPAC)

E. K. Schillinger, S. Shim, and D. E. P. Telenko, Department of Botany and Plant Pathology, Purdue University, West Lafayette, IN 47907-2054

SOYBEAN (*GLYCINE MAX* P29A19E)

White mold, *Sclerotinia sclerotiorum*

A trial was established at the Pinney Purdue Agricultural Center (PPAC) in Porter County, Indiana. The experiment was a randomized complete block design with four replications. Plots were 10 feet wide and 30 feet long and consisted of four rows, and the two center rows were used for evaluation. The previous crop was corn. Standard practices for soybean production in Indiana were followed. Soybean cultivar P29A19E was planted in 30-inch row spacing at a rate of 8 seeds/foot on May 22. Fungicides were applied on July 19 at the full bloom (R2) growth stage. Fungicide and biologicals were applied at 15 gal/acre and 40 psi using a Lee self-propelled sprayer equipped with a 10-foot boom, fitted with six TJ-VS 8002 nozzles spaced 20 inches apart. Disease ratings were rated on September 4, but no disease was detected. Stand count was taken at V8 growth stage on July 3. Canopy greenness was visually assessed as a percentage (0–100%) of crop canopy green. The two center rows of each plot were harvested on October 2, and yields were adjusted to 13% moisture. All data were analyzed in SAS 9.4 (SAS Institute, Cary, NC). A generalized linear mixed model analysis of variance was performed using PROC GLIMMIX. Values are least squares means, and values with different letters are significantly different based on Fisher's least significant difference (α = 0.05).

In 2024, weather conditions were not favorable for the diseases. No disease and no phytotoxicity were detected in trial. There was no significant effect of treatment on stand count (Table 39). There was no significant effect of treatment on canopy greenness, harvest moisture, test weight, and yield of soybean.

TABLE 39. *Effect of Treatment on Stand Count, Canopy Greenness, and Yield of Soybean*

TREATMENT AND RATE/ACRE[z]	STAND COUNT[y] #/ACRE	CANOPY GREEN[x] %	HARVEST MOISTURE %	TEST WEIGHT LB/BU	YIELD[w] BU/ACRE
Nontreated control	120,921	9.3	13.9	56.6	53.6
Approach 8.0 fl oz	126,150	18.0	14.0	55.5	59.7
Double Nickel 2.0 qt	125,496	5.5	13.8	56.1	55.8
Serifel 16.0 fl oz	118,742	17.0	13.9	55.6	56.7
Actinovate 12.0 oz	124,843	12.5	13.8	55.5	60.0
BotryStop 2.0 lb	128,329	9.3	13.8	58.3	54.3
P-value[v]	0.2974	0.2840	0.8489	0.2744	0.4095

[z] Fungicide and biologicals were applied on July 19 at the R2 growth stage.

[y] Stand count was taken at V8 growth stage on July 3.

[x] Canopy greenness was visually assessed as a percentage (0–100%) of crop canopy green on September 16.

[w] Yields were adjusted to 13% moisture and harvested on October 2.

[v] All data were analyzed in SAS 9.4 (SAS Institute, Cary, NC). A generalized linear mixed model analysis of variance was performed using PROC GLIMMIX. Values are least squares means, and values with different letters are significantly different based on Fisher's least significant difference (α = 0.05).

FUNGICIDE EVALUATION FOR WHITE MOLD IN SOYBEAN IN NORTHWESTERN INDIANA, 2024 (SOY24-31.PPAC)

C. Rocco da Silva, S. Shim, and D. E. P. Telenko, Department of Botany and Plant Pathology, Purdue University, West Lafayette, IN 47907-2054

SOYBEAN (*GLYCINE MAX* P29A19E)

White mold, *Sclerotinia sclerotiorum*

A trial was established at the Pinney Purdue Agricultural Center (PPAC) in Porter County, Indiana. The experiment was a randomized complete block design with four replications. Plots were 10 feet wide and 30 feet long and consisted of four rows, and the two center rows were used for evaluation. The previous crop was corn. Standard practices for soybean production in Indiana were followed. Soybean cultivar P29A19E was planted in 30-inch row spacing at a rate of 8 seeds/foot on May 22. *Sclerotinia sclerotiorum* infested sorghum 1.25 g/foot was applied in-furrow at planting. Fungicides were applied on July 12 at the beginning bloom (R1); July 19 seven days after application (DAA), at full bloom (R2) and July 25, 12 DAA, at the beginning pod (R3) growth stages. Fungicides were applied at 15 gal/acre and 40 psi using a CO_2 backpack sprayer at R1 and a Lee self-propelled sprayer equipped with a 10-foot boom, fitted with six TJ-VS 8002 nozzles spaced 20 inches apart, at 5 mph at R2 and R3. Disease ratings were rated on September 4 at full seed (R6) growth stage. White mold disease incidence was assessed by counting the number of plants out of 30 in each plot with symptoms. For disease severity, each plant that was observed was rated according to the following disease category: 0 = no disease, 1 = lateral branches with white mycelium and lesions, 2 = main stem with white mycelium and sclerotia present, 3 = entire plant wilted/plant death. White mold disease severity index: DSI = Total of severity score (0–3) from 30 plants/0.9. Canopy greenness was visually assessed as a percentage (0–100%) of crop canopy green on September 16. The two center rows of each plot were harvested on October 2, and yields were adjusted to 13% moisture. All data were analyzed in SAS 9.4 (SAS Institute, Cary, NC). A generalized linear mixed model analysis of variance was performed using PROC GLIMMIX. Values are least squares means, and values with different letters are significantly different based on Fisher's least significant difference (α = 0.05).

In 2024, weather conditions were not favorable for the disease. White mold was the most prominent disease in the trial but only reached low severity. No significant differences were detected between fungicide programs and nontreated control for white mold DI and DSI (Table 40). There was no significant effect on the percent canopy greenness, harvest moisture, test weight, and yield of soybean.

TABLE 40. *Effect of Treatment on White Mold, Canopy Greenness, and Yield of Soybean*

TREATMENT, RATE/ACRE, AND TIMING[z]	WHITE MOLD DI[y]	WHITE MOLD DSI[x]	CANOPY % GREEN[w]	HARVEST MOISTURE %	TEST WEIGHT LB/BU	YIELD[v] BU/ACRE
Nontreated control	0.5	1.4	3.0	14.2	56.1	54.6
Domark 5.0 fl oz at R1 fb Domark 5.0 fl oz at 7 DAA	1.0	3.1	5.5	14.2	55.9	52.7
Domark 5.0 fl oz at R1 fb Domark 5.0 fl oz at 12 DAA	0.3	0.6	1.5	13.7	54.5	52.1
Affiance 10.0 fl oz at R1 fb Affiance 10.0 fl oz at 7 DAA	0.3	0.6	6.3	14.5	55.0	51.8
Affiance 10.0 fl oz at R1 fb Affiance 10.0 fl oz at 12 DAA	0.0	0.0	1.0	14.3	57.1	57.0
Affiance 14.0 fl oz at R1 fb Affiance 14.0 fl oz at 12 DAA	0.5	1.4	1.8	14.1	55.4	56.1
Domark 5.0 fl oz + Topsin 20 fl oz at R1 fb Domark 5.0 fl oz + Topsin 20 fl oz at 12 DAA	0.3	0.8	0.8	14.1	55.8	57.8
Affiance 10.0 fl oz at R1 fb Endura 7.5 oz at 12 DAA	0.5	1.4	1.3	14.0	53.2	53.2
P-value[u]	*0.5731*	*0.4375*	*0.1662*	*0.2858*	*0.5006*	*0.6592*

[z] Fungicides were applied on July 12 at the beginning bloom (R1), on July 19 7 days after application (DAA) at full bloom (R2), and on July 25 12 DAA at the beginning pod (R3) growth stages. fb = followed by.

[y] White mold disease incidence was assessed by counting the number of plants out of 30 in each plot with symptoms on September 4 at full seed (R6) growth stage.

[x] White mold disease severity index: DSI = Total of severity score (0–3) from 30 plants/0.9.

[w] Canopy greenness was visually assessed as a percentage (0–100%) of crop canopy green on September 16.

[v] Yields were adjusted to 13% moisture and harvested on October 2.

[u] All data were analyzed in SAS 9.4 (SAS Institute, Cary, NC). A generalized linear mixed model analysis of variance was performed using PROC GLIMMIX. Values are least squares means, and values with different letters are significantly different based on Fisher's least significant difference (α = 0.05).

SOUTHWEST PURDUE AGRICULTURAL CENTER (SWPAC)

FUNGICIDE COMPARISON FOR FOLIAR DISEASE IN CORN IN SOUTHWESTERN INDIANA, 2024 (COR24-11.SWPAC)

E. S. Peña, K. M. Goodnight, and D. E. P. Telenko, Department of Botany and Plant Pathology, Purdue University, West Lafayette, IN 47907-2054

CORN (*ZEA MAYS* P0574AMXT)

Gray leaf spot *Cercospora zeae-maydis*
Southern rust, *Puccinia polysora*
Tar spot, *Phyllachora maydis*

A trial was established at the Southwest Purdue Agricultural Center (SWPAC) in Knox County, Indiana. The experiment was a randomized complete block design with four replications. Plots were 10 feet wide and 30 feet long and consisted of four rows, and the two center rows were used for evaluation. The previous crop was soybean. Standard practices for grain corn production in Indiana were followed. Corn hybrid P0574AMXT was planted in 30-inch row spacing at a rate of 27,000 seeds/acre on May 11. Fungicide applications were applied on 15 July at tassel/silk (VT/R1) growth stage. All foliar fungicide applications were applied at 15 gal/acre and 40 psi using a Lee self-propelled sprayer equipped with a 10-foot boom, fitted with six TJ-VS 8002 nozzles spaced 20 inches apart. Disease ratings were assessed on August 19 at dent (R5) growth stage. Tar spot, gray leaf spot (GLS), and southern rust were rated for disease severity by visually assessing the percentage of affected leaf area (0–100%) at the ear leaf on five plants per plot and averaged before analysis. The two center rows of each plot were harvested on October 7, and yields were adjusted to 15.5% moisture. All data were analyzed in SAS 9.4 (SAS Institute, Cary, NC). A generalized linear mixed model analysis of variance was performed using PROC GLIMMIX. Values are least squares means, and values with different letters are significantly different based on Fisher's least significant difference (α = 0.05).

In 2024, weather conditions were not favorable for disease development. Tar spot, GLS, and southern rust were present in the trial but only reached low levels. All fungicide treatments reduced the total foliar disease (tar spot, GLS, and southern rust) as compared to the nontreated control except Miravis Neo (Table 41). No significant differences between treatments were observed for harvest moisture, test weight, and yield of corn.

TABLE 41. *Effect of Treatment on Foliar Disease Severity and Yield of Corn*

TREATMENT AND RATE/ACRE[z]	FOLIAR DISEASE SEVERITY[y] %	HARVEST MOISTURE %	TEST WEIGHT LB/BU	YIELD[x] BU/ACRE
Nontreated control	2.4 a	13.7	17.5	182.3
Veltyma 3.34 SC 7.0 fl oz	0.3 b	14.2	20.0	178.0
Delaro Complete 3.82 SC 8.0 fl oz	0.8 b	13.9	15.0	190.7
Aproach Prima 2.34 SC 6.8 fl oz	0.9 b	14.3	7.5	194.1
Adastrio 4.0 SC 8.0 fl oz	0.7 b	14.0	0.0	190.4
Miravis Neo 2.5 EC 13.7 fl oz	2.0 a	13.8	10.0	185.1
Trivapro 2.21 SE 13.7 fl oz	0.4 b	13.9	32.5	179.4
Headline AMP 1.68 SC 10.0 fl oz	0.4 b	13.9	17.5	189.6
Proline 480 SC 5.7 fl oz	1.0 b	14.0	25.0	187.9
Quadris 2.08 SC 6.0 fl oz	0.9 b	13.7	27.5	184.3
Tilt 3.6 EG 4.0 fl oz	0.7 b	14.0	20.0	191.6
P-value[w]	0.0003	0.8982	0.2198	0.3062

[z] Fungicides were applied on July 15 at tassel/silk (VT/R1) growth stage.

[y] Foliar diseases were visually assessed as a percentage (0–100%) of affected leaf area on five plants in each plot at the ear leaf on August 19 at dent (R5) growth stage (included tar spot, gray leaf spot and southern rust).

[x] Yields were adjusted to 15.5% moisture and harvested on October 7.

[w] All data were analyzed in SAS 9.4 (SAS Institute, Cary, NC). A generalized linear mixed model analysis of variance was performed using PROC GLIMMIX. Values are least squares means, and values with different letters are significantly different based on Fisher's least significant difference ($\alpha = 0.05$).

FUNGICIDE COMPARISON FOR FOLIAR DISEASE IN CORN IN SOUTHWESTERN INDIANA, 2024 (CORN 24-30.SWPAC)

E. Peña Roncancio, S. Shim, and D. E. P. Telenko, Department of Botany and Plant Pathology, Purdue University, West Lafayette, IN 47907-2054

CORN (*ZEA MAYS* P0574AM)

Tar spot, *Phyllachora maydis*
Gray leaf spot, *Cercospora zeae-maydis*
Southern rust, *Puccinia polysora*

A trial was established at the Southwest Purdue Agricultural Center (SWPAC) in Knox County, Indiana. The experiment was a randomized complete block design with four replications. Plots were 10 feet wide and 30 feet long and consisted of four rows, and the two center rows were used for evaluation. The previous crop was soybean. Standard practices for grain corn production in Indiana were followed. Corn hybrid P0574AMXT was planted in 30-inch row spacing at a rate of 27,000 seeds/acre on May 11. All fungicides were applied at 15 gal/acre and 40 psi using either a CO_2 backpack sprayer on (V9) or a Lee self-propelled sprayer on (VT/R1) equipped with a 10-foot boom, fitted with six TJ-VS 8002 nozzles spaced 20 inches apart. Fungicides were applied on June 27 at V9 and July 15 at the tassel/silk (VT/R1) growth stages. Disease ratings were assessed on August 19 at dent (R5) growth stage. Tar spot, gray leaf spot (GLS), and southern rust were rated for severity by visually assessing the percentage of symptomatic leaf area (0–100%) per leaf on five plants in each plot at the ear leaf. Values for the five leaves were averaged before analysis. The two center rows of each plot were harvested on October 7, and yields were adjusted to 15.5% moisture. All data were analyzed in SAS 9.4 (SAS Institute, Cary, NC). A generalized linear mixed model analysis of variance was performed using PROC GLIMMIX. Values are least squares means, and values with different letters are significantly different based on Fisher's least significant difference ($\alpha = 0.05$).

In 2024, weather conditions were not favorable for disease development. Tar spot, GLS, and southern rust were detected in plots but reached low severity. Tar spot was significantly reduced by Veltyma at R1, Topguard at V9 followed by Adastrio at R1, Adastrio at V9 followed by Topguard at R1, and Delaro Complete + OR-99EPA at R1 as compared to the nontreated control (Table 42). All fungicide programs significantly reduced southern rust as compared to nontreated control. There was no significant effect of treatment on GLS severity, harvest moisture, test weight, and yield of corn.

TABLE 42. *Effect of Treatment on Foliar Disease and Yield of Corn*

TREATMENT, RATE/ACRE, AND TIMING[z]	TAR SPOT[y] % SEVERITY	GLS[y] % SEVERITY	SR[y] % SEVERITY	HARVEST MOISTURE %	TEST WEIGHT LB/BU	YIELD[x] BU/ACRE
Nontreated control	0.3 a	0.1	0.5 a	14.2	57.8	189.4
Veltyma 7.0 fl oz + NIS 0.25% V/V at R1	0.2 bc	0.1	0.1 b	14.1	57.8	182.1
Delaro Complete 8.0 fl oz + NIS 0.25% V/V at R1	0.3 ab	0.1	0.1 b	14.3	57.9	189.1
Topguard 8.0 fl oz at V9 fb Adastrio 8.0 fl oz + NIS 0.25% V/V at R1	0.1 c	0.0	0.1 b	14.3	57.8	195.2
Adastrio 8.0 fl oz at V9 fb Topguard EQ 7.0 fl oz + NIS 0.25% V/V at R1	0.2 bc	0.0	0.1 b	14.3	57.8	198.2
Veltyma 7.0 fl oz + OR-099EPA 0.4% v/v at R1	0.3 a	0.1	0.2 b	14.2	58.5	184.8
Delaro Complete 8.0 fl oz + OR-099EPA 0.4% v/v at R1	0.1 c	0.1	0.1 b	14.1	57.7	193.5
P-value[w]	*0.0094*	*0.0855*	*0.0001*	*0.9518*	*0.6806*	*0.1258*

[z] Fungicides were applied on June 27 at V9 and July 15 at the tassel/silk (VT/R1) growth stages; fb = followed by.

[y] Foliar disease severity was visually assessed as a percentage (0–100%) of affected leaf area on five plants in each plot at the ear leaf on August 19 at dent (R5) growth stage. GLS = gray leaf spot; SR = southern rust.

[x] Yields were adjusted to 15.5% moisture and harvested on October 7.

[w] All data were analyzed in SAS 9.4 (SAS Institute, Cary, NC). A generalized linear mixed model analysis of variance was performed using PROC GLIMMIX. Values are least squares means, and values with different letters are significantly different based on Fisher's least significant difference (α = 0.05).

EVALUATION OF FUNGICIDES FOR FOLIAR DISEASES ON SOYBEAN IN SOUTHWESTERN INDIANA, 2024 (SOY24-02.SWPAC)

S. Shim and D. E. P. Telenko, Department of Botany and Plant Pathology, Purdue University, West Lafayette, IN 47907-2054

SOYBEAN (*GLYCINE MAX* P29A19E)

Frogeye leaf spot, *Cercospora sojina*
Septoria brown spot, *Septoria glycines*

A trial was established at the Southwest Purdue Agricultural Center (SWPAC) in Knox County, Indiana. The experiment was a randomized complete block design with four replications. Plots were 10 feet wide and 30 feet long and consisted of four rows, and the two center rows were used for evaluation. The previous crop was corn. Standard practices for soybean production in Indiana were followed. Soybean cultivar P29A19E was planted in 30-inch row spacing at a rate of 150,000 seeds/acre on May 24. All fungicides were applied at 15 gal/acre and 40 psi using a Lee self-propelled sprayer equipped with a 10-foot boom, fitted with six TJ-VS 8002 nozzles spaced 20 inches apart. Fungicides were applied on July 15 at the beginning pod (R3) growth stage. Foliar disease ratings were rated on August 19 at full seed (R6) growth stage. The two center rows of each plot were harvested on September 11, and yields were adjusted to 13% moisture. All data were analyzed in SAS 9.4 (SAS Institute, Cary, NC). A generalized linear mixed model analysis of variance was performed using PROC GLIMMIX. Values are least squares means, and values with different letters are significantly different based on Fisher's least significant difference ($\alpha = 0.05$).

In 2024, weather conditions were not favorable for disease development. Septoria brown spot and and frogeye leaf spot were present in the trial but only reached low levels. There was no significant effect on harvest moisture, test weight, and yield of soybean (Table 43).

TABLE 43. *Effect of Treatment on Yield of Soybean*

TREATMENT AND RATE/ACRE[z]	HARVEST MOISTURE %	TEST WEIGHT LB/BU	YIELD[y] BU/ACRE
Nontreated control	9.6	60.2	61.6
Topguard EQ 4.29 SC 5.0 fl oz	10.1	55.5	56.6
Lucento 4.17 SC 5.0 fl oz	10.7	61.4	61.1
Trivapro 2.21 SE 13.7 fl oz	10.0	59.5	56.4
Quadris 2.08 SC 6.0 fl oz	10.9	62.4	58.9
Veltyma 3.34 SC 7.0 fl oz	11.2	61.0	60.1
Revytek 4.44 SC 8.0 fl oz	11.2	61.0	61.5
Echo 2.21 SE 36.0 fl oz + Folicur 3.6 FL 4.0 fl oz + Topsin 4.5 SC 4.5 fl oz	11.9	60.7	58.7
Delaro Complete 3.82 SC 8.0 fl oz	12.0	60.3	58.3
Miravis Neo 2.5 EC 13.7 fl oz	10.8	61.8	59.9
Topsin 4.5 SC 4.5 20.0 fl oz	10.2	61.1	59.5
P-value[x]	*0.2181*	*0.3463*	*0.5801*

[z] Fungicides were applied on July 15 at the R3 (beginning pod) growth stage, and all treatments contained a nonionic surfactant (Preference) at a rate of 0.25% v/v.

[y] Yields were adjusted to 13% moisture and harvested on September 11.

[x] All data were analyzed in SAS 9.4 (SAS Institute, Cary, NC). A generalized linear mixed model analysis of variance was performed using PROC GLIMMIX. Values are least squares means, and values with different letters are significantly different based on Fisher's least significant difference (α = 0.05).

EVALUATION OF FUNGICIDES FOR FOLIAR DISEASES ON SOYBEAN IN SOUTHWESTERN INDIANA, 2024 (SOY24-21.SWPAC)

S. Shim and D. E. P. Telenko, Department of Botany and Plant Pathology, Purdue University, West Lafayette, IN 47907-2054

SOYBEAN (*GLYCINE MAX* P29A19E)

Septoria brown spot, *Septoria glycines*
Frogeye leaf spot, *Cercospora sojina*

A trial was established at the Southwest Purdue Agricultural Center (SWPAC) in Knox County, Indiana. The experiment was a randomized complete block design with four replications. Plots were 10 feet wide and 30 feet long and consisted of four rows, and the two center rows were used for evaluation. The previous crop was corn. Standard practices for soybean production in Indiana were followed. Soybean cultivar P29A19E was planted in 30-inch row spacing at a rate of 150,000 seeds/acre on May 24. Fungicides were applied on June 27 at V5 and July 15 at the beginning pod (R3) growth stage. All fungicides were applied at 15 gal/acre and 40 psi using either a CO_2 backpack sprayer (V5) or a Lee self-propelled sprayer (R3) equipped with a 10-foot boom, fitted with six TJ-VS 8002 nozzles spaced 20 inches apart. Foliar disease ratings were rated on August 19 at full seed (R6) growth stage. Septoria brown spot (SBS) and frogeye leaf spot (FLS) were rated for disease severity by visually assessing the percentage of symptomatic leaf area in the canopy. The two center rows of each plot were harvested on September 11, and yields were adjusted to 13% moisture. All data were analyzed in SAS 9.4 (SAS Institute, Cary, NC). A generalized linear mixed model analysis of variance was performed using PROC GLIMMIX. Values are least squares means, and values with different letters are significantly different based on Fisher's least significant difference (α = 0.05).

In 2024, weather conditions were not favorable for disease development. SBS and FLS were present in the trial but only reached low levels. All treatments significantly reduced SBS over the nontreated control (Table 44). There was no significant effect of treatment on FLS severity. There was no significant effect on harvest moisture, test weight, and yield of soybean.

TABLE 44. *Effect of Treatment on Foliar Disease and Yield of Soybean*

TREATMENT, RATE/ACRE, AND TIMING[z]	SBS[y] %	FLS[y] %	HARVEST MOISTURE %	TEST WEIGHT LB/BU	YIELD[x] BU/ACRE
Nontreated control	2.2 a	0.2	14.4	59.4	65.2
Lucento 4.17 SC 5.0 fl oz at R3	0.3 b	0.1	10.5	60.9	63.4
Adastrio 4.0 SC 8.0 fl oz at R3	0.3 b	0.1	13.2	59.0	63.5
Topguard EQ 4.29 SC 7.0 fl oz at V4 fb Lucento 4.17 SC 5.0 fl oz at R3	0.3 b	0.1	10.9	60.2	66.6
Topguard EQ 4.29 SC 7.0 fl oz at V4 fb Adastrio 4.0 SC 8.0 fl oz at R3	0.3 b	0.1	11.6	60.4	64.7
Delaro Complete 3.82 SC 8.0 fl oz at R3	0.4 b	0.1	12.3	58.9	66.7
Revytek 4.44 SC 8.0 fl oz at R3	0.2 b	0.1	13.8	59.3	65.1
P-value[w]	*0.0357*	*0.1108*	*0.2010*	*0.8983*	*0.8370*

[z] Fungicides were applied on June 27 at V5 and July 15 at the R3 (beginning pod) growth stage, and all treatments contained a nonionic surfactant (Preference) at a rate of 0.25% v/v. fb = followed by.

[y] Foliar disease severity was rated on a scale of 0–100% of canopy within a plot with disease symptoms on August 19 at full seed (R6) growth stage. SBS = Septoria brown spot; FLS = frogeye leaf spot.

[x] Yields were adjusted to 13% moisture and harvested on September 11.

[w] All data were analyzed in SAS 9.4 (SAS Institute, Cary, NC). A generalized linear mixed model analysis of variance was performed using PROC GLIMMIX. Values are least squares means, and values with different letters are significantly different based on Fisher's least significant difference (α = 0.05).

EVALUATION OF FUNGICIDE EFFICACY FOR FUSARIUM HEAD BLIGHT OF WHEAT IN SOUTHWESTERN INDIANA, 2024 (WHT24-03.SWPAC)

S. Shim and D. E. P. Telenko., Department of Botany and Plant Pathology, Purdue University, West Lafayette, IN 47907-2054

WHEAT (*TRITICUM AESTIVUM* P25R40)

Fusarium head blight, *Fusarium graminearum*

A trial was established at the Southwest Purdue Agricultural Center (SWPAC) in Knox County, Indiana. The experiment was a randomized complete block design with four replications. Plots were 7.5 feet wide and 20 feet long and consisted of 12 rows spaced 7.5 inches apart, and the center of each plot was used for evaluation. The previous crop was corn. Wheat cultivar P25R40 was planted in 7.5-inch row spacing using a drill on October 22, 2023. All fungicide applications were applied at 15 gal/acre and 40 psi using a CO_2 backpack sprayer equipped with a 10-foot boom, fitted with six TJ-VS 8002 nozzles spaced 20 inches apart and directed forward and backward at a 45-degree angle. Fungicides were applied on May 1 at Feekes growth stage 10.5.1 and five days after on May 6. All plots were inoculated with a mixture of isolates of *Fusarium graminearum* endemic to Indiana on May 1 with a spore suspension (50,000 spores/ml) applied at 300 ml/plot. Disease ratings were assessed on May 23. Fusarium head blight (FHB) incidence was measured as the number of infected heads out of 60 plants in each plot and calculated as a percentage. FHB severity was rated by visually assessing the percentage (0–100%) of the infected heads. The FHB index was calculated as % FHB incidence multiplied by average FHB severity /100 per plot. Values for each plot were averaged before analysis. The eight center rows of each plot were harvested with a Kincaid plot combine on June 18, and yields were adjusted to 13.5% moisture for comparison. A subsample of grain was taken from each plot and partitioned for deoxynivalenol (DON) analysis completed by the University of Minnesota DON testing lab and to determine Fusarium damaged kernels (FDK) by visually assessing the percentage (0–100%) of the infected heads. All data were analyzed in SAS 9.4 (SAS Institute, Cary, NC). A generalized linear mixed model analysis of variance was performed using PROC GLIMMIX. Values are least squares means, and values with different letters are significantly different based on Fisher's least significant difference (α = 0.05).

In 2024, weather conditions were favorable for FHB. FHB incidence, severity, and index were significantly reduced by all fungicide applications when compared to the nontreated control except Miravis Era (Table 45). The percent of FDK was significantly reduced by all fungicide programs over the nontreated control except Prosaro 421SC and Sphaerex 2.50SC. The concentration of DON was significantly reduced by all fungicide applications over nontreated control except by Prosaro 421SC. All treatments increased yield over nontreated control except Miravis Era.

TABLE 45. *Effect of Fungicide on Fusarium Head Blight (FHB), Fusarium Damaged Kernels (FDK), DON, and Yield of Wheat*

TREATMENT, RATE/ACRE, AND TIMING[z]	FHB % INCIDENCE[y]	FHB % SEVERITY[x]	FHB INDEX[w]	FDK[v] %	DON[u] PPM	YIELD[t] BU/ACRE
Nontreated control	94.6 a	25.0 a	23.8 a	14.0 d	9.4 a	59.6 c
Prosaro 421SC 6.5 fl oz at 10.5.1	60.4 bc	11.2 bc	7.2 bc	9.5 ab	3.6 bc	70.6 a
Miravis Era 10.2 fl oz at 10.5.1	90.8 a	21.0 a	19.3 a	14.5 d	9.1 a	61.2 bc
Miravis Ace 5.2SC 13.7 fl oz at 10.5.1	46.3 bc	9.9 bc	4.6 bc	12.0 bc	3.7 bc	67.2 ab
Prosaro Pro 400SC 10.3 fl oz at 10.5.1	62.9 b	12.6 bc	8.2 bc	10.0 bc	4.3 b	68.2 ab
Sphaerex 2.50SC 7.3 fl oz at 10.5.1	40.0 c	8.8 c	3.5 c	9.3 bc	3.9 bc	71.0 a
Miravis Ace 5.2SC 13.7 fl oz at 10.5.1 fb Prosaro Pro 400SC 10.3 fl oz at 10.5.1 + 5 d	65.4 b	13.7 b	9.2 b	14.5 cd	2.5 bc	68.6 a
Miravis Ace 5.2SC 13.7 fl oz at 10.5.1 fb Sphaerex 2.50SC 7.3 fl oz at 10.5.1 + 5 d	55.8 bc	11.2 bc	6.5 bc	10.8 a	2.1 c	72.8 a
Miravis Ace 5.2SC 13.7 fl oz at 10.5.1 fb Tebuconazole 4.0 fl oz at 10.5.1 + 5 d	55.8 bc	10.3 bc	6.3 bc	12.8 ab	3.0 bc	9.2 a
P-value[s]	0.0002	0.0001	0.0001	0.0001	0.0001	0.0131

[z] Fungicides were applied on May 1 at Feekes growth stages 10.5.1 and 10.5.10 + 5 days after on May 6. All treatments contained a nonionic surfactant (Preference) at a rate of 0.125% v/v. All plots were inoculated with a mixture of isolates of *Fusarium graminearum* endemic to Indiana on May 1 with a spore suspension (50,000 spores/ml) applied at 300 ml/plot. fb = followed by.

[y] FHB incidence was measured as the number of infected heads out of 60 plants in each plot and calculated as a percentage on May 23.

[x] FHB severity was rated by visually assessing the percentage of the infected head.

[w] The FHB index was calculated as % FHB incidence multiplied by average FHB severity/100 per plot.

[v] Visual assessment of the percentage of Fusarium damaged kernels (FDK) was performed on July 10.

[u] Analysis of the mycotoxin DON was completed by the University of Minnesota DON Testing Lab.

[t] Yields were adjusted to 13.5% moisture and harvested on June 18.

[s] All data were analyzed in SAS 9.4 (SAS Institute, Cary, NC). A generalized linear mixed model analysis of variance was performed using PROC GLIMMIX. Values are least squares means, and values with different letters are significantly different based on Fisher's least significant difference (α = 0.05).

EVALUATION OF CULTIVARS AND FUNGICIDES FUSARIUM HEAD BLIGHT OF WHEAT IN SOUTHWESTERN INDIANA (WHT24-04_SCAB.SWPAC)

S. Shim and D. E. P. Telenko., Department of Botany and Plant Pathology, Purdue University, West Lafayette, IN 47907-2054

WHEAT (*TRITICUM AESTIVUM* P25R40 AND P25R61)

Fusarium head blight, *Fusarium graminearum*

A trial was established at the Southwest Purdue Agricultural Center (SWPAC) in Knox County, Indiana. The experiment was a strip-plot design with four replications. Plots were 7.5 feet wide and 20 feet long and consisted of 12 rows spaced 7.5 inches apart, and the center of each plot was used for evaluation. The previous crop was corn. On October 11, 2023, wheat cultivars P25R40 and P25R61 were drilled at 7.5-inch spacing. Fungicides were applied on May 1 at Feekes growth stage 10.5.1. All fungicide applications were applied at 15 gal/acre and 40 psi using a CO_2 backpack sprayer equipped with a 10-foot boom, fitted with six TJ-VS 8002 nozzles spaced 20 inches apart and directed forward and backward at a 45-degree angle. All plots were inoculated with a mixture of isolates of *Fusarium graminearum* endemic to Indiana on May 1. Disease ratings were assessed on May 23. Fusarium head blight (FHB) incidence was measured as the number of infected heads out of 60 plants in each plot and calculated as a percentage (0–100%). FHB severity was rated by visually assessing the percentage (0–100%) of the infected head. The FHB index was calculated as % FHB incidence multiplied by average FHB severity/100 per plot. Values for each plot were averaged before analysis. The eight center rows of each plot were harvested with a Kincaid 8XP combine on June 18, and yields were adjusted to 13.5% moisture. A subsample of grain was taken from each plot and partitioned for deoxynivalenol (DON) analysis completed by the University of Minnesota DON testing lab and to determine Fusarium damaged kernels (FDK) by visually assessing the percentage (0–100%) of the infected heads. All data were analyzed in SAS 9.4 (SAS Institute, Cary, NC). A generalized linear mixed model analysis of variance was performed using PROC GLIMMIX. Values are least squares means, and values with different letters are significantly different based on Fisher's least significant difference ($\alpha = 0.05$).

In 2024, weather conditions were favorable for FHB. FHB was the most prominent disease in the trial. The FHB index, FDK, and DON were lowest in the resistant cultivar P25R61 (Table 46). There was no difference between cultivars for moisture, test weight and yield. The FHB index was reduced by all fungicide treatments over nontreated controls. Applications of Miravis Ace had the highest percent of FDK. The concentration of DON was significantly reduced by all the fungicides over the nontreated controls. There was no difference in treatment for moisture, test weight and yield of wheat.

TABLE 46. *Effect of Cultivar and Fungicide on Fusarium Head Blight (FHB), Fusarium Damaged Kernels (FDK), DON, and Yield of Wheat*

TREATMENT AND RATE/ACRE[z]	FHB INDEX[y]	FDK[x] %	DON[w] PPM	MOISTURE %	TEST WEIGHT LB/BU	YIELD[v] BU/ACRE
Cultivar						
P25R40 (scab susceptible)	19.1 a	9.0 a	5.3 a	18.3	51.7	66.3
P25R61 (scab resistant)	5.7 b	6.9 b	1.5 b	18.4	51.4	64.7
Fungicide						
Nontreated control, inoculated control	19.8 a	6.5 c	5.2 a	18.2	51.6	65.7
Nontreated, non-inoculated control	18.9 a	9.3 b	5.4 a	18.6	51.7	67.8
Prosaro 421SC 6.5 fl oz	10.6 b	5.4 c	2.7 b	18.4	51.5	65.3
Miravis Ace 5.2SC 13.7 fl oz	8.3 b	12.3 a	2.6 b	18.2	51.6	63.9
Prosaro Pro 400SC 10.3 fl oz	9.7 b	7.3 bc	2.4 b	18.7	51.7	65.2
Sphaerex 2.50SC 7.3 fl oz	7.1 b	7.1 bc	2.0 b	18.3	51.3	65.2
P-value cultivar[u]	0.0001	0.0088	0.0001	0.7559	0.5915	0.3200
P-value fungicide	0.0001	0.0001	0.0001	0.5142	0.9930	0.8078
P-value cultivar*fungicide	0.2143	0.0001	0.0022	0.3741	0.9727	0.7003

[z] Fungicide treatments were applied on May 1 at Feekes growth stage 10.5.1. All fungicide treatments contained a nonionic surfactant (Preference) at a rate of 0.125% v/v. All plots were inoculated with *Fusarium graminearum* spore suspension (50,000 spores/ml) after the treatment at Feekes 10.5.1. Spore suspension applied at 300 ml/plot with handheld sprayer on May 1.

[y] The FHB index was calculated as FHB incidence multiplied by average FHB severity/100 per plot.

[x] Visual assessment of percentage of Fusarium damaged kernels (FDK) was performed on a subset of grain.

[w] Analysis of the mycotoxin DON was completed by the University of Minnesota DON Testing Lab.

[v] Yields were adjusted to 13.5% moisture and harvested on June 18.

[u] All data were analyzed in SAS 9.4 (SAS Institute, Cary, NC). A generalized linear mixed model analysis of variance was performed using PROC GLIMMIX. Values are least squares means, and values with different letters are significantly different based on Fisher's least significant difference (α = 0.05).

DAVIS PURDUE AGRICULTURAL CENTER (DPAC)

EVALUATION OF DRONE VERSUS GROUND FUNGICIDE APPLICATION METHODS IN CORN IN CENTRAL INDIANA, 2024 (COR24-08.DPAC)

M. S. Mizuno, H. Medenwald, and D. E. P. Telenko, Department of Botany and Plant Pathology, Purdue University, West Lafayette, IN 47907-2054

CORN (ZEA MAYS P0574AM)

Tar spot, *Phyllachora maydis*
Northern corn leaf blight, *Exserohilum turcicum*

A trial was established at the Davis Purdue Agricultural Center (DPAC) in Randolph County, Indiana. The experiment was a randomized complete block design with five replications. Plots were 30 feet wide and 380 feet long and consisted of 12 rows, and the two center rows were used for evaluation. The previous crop was soybean. Standard practices for grain corn production in Indiana were followed. Corn hybrid P0574AM was planted in 30-inch row spacing at a rate of 32,000 seeds/acre on May 7. Veltyma 3.34 S 7.0 fl oz/acre was applied on July 26 at blister (R2) growth stage using two different applicators: a Patriot sprayer equipped with a 30-foot boom, fitted with 18 AIC110006 nozzles spaced 20 inches apart, at 10 mph, and a DJI Agras T30 drone with spray pattern using 16 XRTTeeJet 11001VS nozzles at 10-foot altitude at 15 mph to apply at 2 gal/acre and at 8 mph to apply at 5 gal/acre. Disease ratings were assessed on September 23 at maturity (R6) growth stage. Tar spot stromata severity and northern corn leaf blight (NCLB) were visually assessed as a percentage (0–100%) of symptomatic leaf area on five plants per plot in three locations in each plot and averaged before analysis. Percent canopy greenness was rated by visually assessing the percentage (0–100%) of canopy green on September 23 at maturity (R6) growth stage. The trial was harvested on November 7, and yields were adjusted to 15.5% moisture. All data were analyzed in SAS 9.4 (SAS Institute, Cary, NC). A generalized linear mixed model analysis of variance was performed using PROC GLIMMIX. Values are least squares means, and values with different letters are significantly different based on Fisher's least significant difference (α = 0.05).

In 2024, weather conditions were moderately favorable for disease. Tar spot and NCLB were the most prominent diseases in the trial and reached moderate severity. Tar spot stromata severity was significantly reduced over the nontreated control by all application methods (Table 47). The drone application at 5 gal/acre had the lowest level of tar spot but was not significantly different from the drone application at 2 gal/acre. All applications significantly reduced NCLB severity over the nontreated control, but there was no difference between application methods. There was no significant effect of treatment for canopy greenness, harvest moisture, and yield of corn.

TABLE 47. *Effect of Different Application Methods on Foliar Disease Severity, Canopy Greenness, and Yield Corn*

APPLICATION EQUIPMENT AND GPA[z]	TAR SPOT[y] %	NCLB[y] %	CANOPY GREEN[x] %	HARVEST MOISTURE %	YIELD[w] BU/ACRE
Nontreated control	15.4 a	5.9 a	9.0	15.2	199.4
DJI Agras T30 Drone at 2 GPA	6.6 bc	0.1 b	19.7	15.1	204.4
DJI Agras T30 Drone at 5 GPA	3.1 c	0.0 b	20.6	15.2	203.3
Ground-rig with at 20 GPA	8.1 b	1.4 b	14.0	15.2	201.9
P-value[v]	0.0007	0.0085	0.4102	0.2120	0.2630

[z] Fungicide applications were made on July 26 at blister (R2) growth stage, Veltyma 3.34 S at 7.0 fl oz/acre, and contained a nonionic surfactant (Preference) at a rate of 0.25% v/v. A blue tracer dye was used at 0.25% v/v. GPA = gallons per acre.

[y] Foliar disease severity was visually assessed as a percentage (0–100%) of leaf area on five plants in each plot at the ear leaf on September 23 at maturity (R6) growth stage. NCLB = northern corn leaf blight.

[x] Canopy greenness was visually assessed as a percentage (0–100%) of canopy green on September 23.

[w] Yields were adjusted to 15.5% moisture and harvested on November 7.

[v] All data were analyzed in SAS 9.4 (SAS Institute, Cary, NC). A generalized linear mixed model analysis of variance was performed using PROC GLIMMIX. Values are least squares means, and values with different letters are significantly different based on Fisher's least significant difference (α = 0.05).

EVALUATION OF DRONE VERSUS GROUND FUNGICIDE APPLICATION METHODS IN SOYBEAN IN CENTRAL INDIANA, 2024 (SOY24-05.DPAC)

M. S. Mizuno, H. Medenwald, A. Helms, and D. E. P. Telenko,
Department of Botany and Plant Pathology, Agricultural Research and Graduate Education,
Purdue University, West Lafayette, IN 47907-2054

SOYBEAN (*GLYCINE MAX* P29A19E)

Septoria brown spot, *Septoria glycines*
Downy mildew, *Peronospora manshurica*
Frogeye leaf spot, *Cercospora sojina*

A trial was established at the Davis Purdue Agricultural Center (DPAC) in Randolph County, Indiana. The experiment was a randomized complete block design with three replications. Plots were 30 feet wide and 594 feet long and consisted of 24 rows, and the two center rows were used for evaluation. The previous crop was corn. Standard practices for soybean production in Indiana were followed. Soybean cultivar P29A19E was planted in 15-inch row spacing at a rate of 140,000 seeds/acre on May 14. Delaro Complete 3.82 SC 8.0 fl oz/acre was applied on July 26 at beginning pod (R3) and August 13 at beginning seed (R5) growth stages using two different applicators: a Patriot sprayer equipped with a 30-foot boom, fitted with 18 AIC110006 nozzles spaced 20 inches apart, at 10 mph, and a DJI Agras T30 drone with spray pattern using 16 XRTeeJet 11001VS nozzles at 10-foot altitude at 15 mph to apply at 2 gal/acre and at 8 mph to apply at 5 gal/acre. Disease ratings were assessed on September 12 at beginning maturity (R7) growth stage. Frogeye leaf spot (FLS) and downy mildew (DM) were rated in the upper canopy, and Septoria brown spot (SBS) was rated in the lower canopy. Disease severity of each disease was visually assessed by as a percentage (0–100%) of symptomatic leaf area in three locations in each plot. All ratings were averaged in each plot before analysis. Soybean plots were harvested on October 9, and yields were adjusted to 13% moisture. All data were analyzed in SAS 9.4 (SAS Institute, Cary, NC). A generalized linear mixed model analysis of variance was performed using PROC GLIMMIX. Values are least squares means, and values with different letters are significantly different based on Fisher's least significant difference (α = 0.05).

In 2024, weather conditions were not favorable for disease. SBS and FLS were the most prominent diseases in the trial and reached low severity. FLS severity was significantly reduced over the nontreated control by all treatments at R3 and R5, but there was no difference between application method or application time (Table 48). There was no significant effect of treatment on DM and SBS severity. There was no significant difference between application method and nontreated control for harvest moisture and yield of soybean.

TABLE 48. *Effect of Different Application Methods on Disease Severity and Yield of Soybean*

APPLICATION EQUIPMENT, GPA, AND TIMING[z]	FLS[y] %	SBS[y] %	DM[y] %	MOISTURE %	YIELD[x] BU/ACRE
Nontreated control	2.2 a	2.6	0.8	9.7	70.3
DJI Agras T30 Drone 2 GPA at R3	0.5 b	0.7	0.5	9.7	68.5
DJI Agras T30 Drone 5 GPA at R3	0.1 b	1.0	0.3	9.7	69.9
Ground-rig 20 GPA at R3	0.3 b	0.7	0.3	9.7	67.2
DJI Agras T30 Drone 2 GPA at R5	0.4 b	0.9	0.3	9.7	69.2
DJI Agras T30 Drone 5 GPA at R5	0.1 b	0.6	0.3	9.7	68.5
Ground-rig 20 GPA at R5	0.4 b	0.7	0.4	9.7	65.0
P-value[w]	*0.0399*	*0.1269*	*0.4190*	*0.4481*	*0.0895*

[z] Fungicide applications were made on July 26 at beginning pod (R3) and August 13 at beginning seed (R5) growth stages of Delaro Complete 3.82 SC 8.0 fl oz/acre and contained a nonionic surfactant (Preference) at a rate of 0.25% v/v. A blue tracer dye was used at 0.25% v/v. GPA = gallons per acre.

[y] Foliar disease incidence was rated on scale of 0–100% of plants with disease symptoms on September 12 at full seed (R6) growth stage. FLS = frogeye leaf spot in upper canopy; SBS = Septoria brown spot in lower canopy; DM = downy mildew in upper canopy.

[x] Yields were adjusted to 13% moisture and harvested on October 9.

[w] All data were analyzed in SAS 9.4 (SAS Institute, Cary, NC). A generalized linear mixed model analysis of variance was performed using PROC GLIMMIX. Values are least squares means, and values with different letters are significantly different based on Fisher's least significant difference (α = 0.05).

NORTHEAST PURDUE AGRICULTURAL CENTER (NEPAC)

EVALUATION OF DRONE VERSUS GROUND FUNGICIDE APPLICATION METHODS IN CORN IN NORTHEAST INDIANA, 2024 (COR24-09.NEPAC)

M. S. Mizuno, H. Medenwald, and D. E. P. Telenko, Department of Botany and Plant Pathology, Purdue University, West Lafayette, IN 47907-2054

CORN (ZEA MAYS P0574AM)

Tar spot, *Phyllachora maydis*
Northern corn leaf blight, *Exserohilum turcicum*

A trial was established at the Northeast Purdue Agricultural Center (NEPAC) in Whitley County, Indiana. The experiment was a randomized complete block design with six replications. Plots were 30 feet wide and 360 feet long and consisted of 12 rows, and the two center rows were used for evaluation. The previous crop was soybean. Standard practices for grain corn production in Indiana were followed. Corn hybrid P0574AM was planted in 30-inch row spacing at a rate of 32,000 seeds/acre on May 30. Veltyma 3.34 S 7.0 fl oz/acre was applied on August 8 at milk (R3) growth stage using two different applicators: a Case IH 2250 Patriot sprayer equipped with a 30-foot boom, fitted with 18 AITTJ60-11008VP nozzles spaced 20 inches apart, at 8 mph and applied at 15 gal/acre and 60 psi and a DJI Agras T30 drone with spray pattern using 16 XRTTeeJet 11001VS nozzles at 10-foot altitude at 15 mph applied at 2 gal/acre and at 8 mph to apply at 5 gal/acre. Disease rating was assessed on September 19 at dent (R5) growth stage. Tar spot stromata severity and northern corn leaf blight (NCLB) were visually assessed as a percentage (0–100%) of symptomatic leaf area on five plants per plot at three locations in each plot and averaged before analysis. Percent canopy greenness was rated by visually assessing the percentage (0–100%) of canopy green on October 2 at dent (R5) growth stage. The trial was harvested on October 24, and yields were adjusted to 15.5% moisture. Data were averaged before analysis and subjected to mixed model analysis of variance in SAS 9.4 (SAS Institute, Cary, NC). A generalized linear mixed model analysis of variance was performed using PROC GLIMMIX. Values are least squares means, and values with different letters are significantly different based on Fisher's least significant difference (α = 0.05).

In 2024, weather conditions were favorable for disease. Tar spot and NCLB were the most prominent disease in the trial and reached low severity. The fungicide sprayed with the ground rig significantly reduced tar spot stromata severity over the drone at 5 gal/acre, but was not significant from the nontreated control (Table 49). There was no significant effect of application type for NCLB severity. There were no significant differences in treatments for percentage of canopy greenness. Veltyma sprayed with the ground rig resulted in significantly higher harvest moisture over the nontreated control. There was no significant difference between application methods of fungicide and nontreated control for yield of corn.

TABLE 49. *Effect of Different Application Methods on Foliar Disease Severity, Canopy Greenness, and Yield of Corn*

APPLICATION EQUIPMENT AND GPA[z]	TAR SPOT STROMATA[y] %	NCLB SEVERIT[y] %	CANOPY GREEN[x] %	HARVEST MOISTURE %	YIELD[w] BU/ACRE
Nontreated control	0.3 ab	0.6	40.8	18.3 b	185.5
DJI Agras T30 Drone at 2 GPA	0.4 ab	0.4	51.1	18.4 b	181.4
DJI Agras T30 Drone at 5 GPA	0.6 a	0.2	40.3	18.2 b	182.7
Ground-rig at 20 GPA	0.0 b	0.1	52.2	19.0 a	185.0
P-value[v]	0.0223	0.2668	0.0699	0.0075	0.5980

[z] Fungicide treatment was applied on August 8 at milk (R3) growth stage. All foliar treatments contained Veltyma 3.34 S at 7.0 fl oz/acre and a nonionic surfactant (Preference) at a rate of 0.25% v/v. A blue tracer dye was used at 0.25% v/v. GPA = gallons per acre.

[y] Foliar disease severity was visually assessed as a percentage (0–100%) of leaf area on five plants in each plot at the ear leaf on 19 September at dent (R5) growth stage. NCLB = northern corn leaf blight.

[x] Canopy greenness was visually assessed as a percentage (0–100%) of canopy green on October 2.

[w] Yields were adjusted to 15.5% moisture and harvested on October 24.

[v] All data were analyzed in SAS 9.4 (SAS Institute, Cary, NC). A generalized linear mixed model analysis of variance was performed using PROC GLIMMIX. Values are least squares means, and values with different letters are significantly different based on Fisher's least significant difference (α = 0.05).

EVALUATION OF DRONE VERSUS GROUND FUNGICIDE APPLICATION METHODS IN SOYBEAN IN NORTHEAST INDIANA, 2024 (SOY24-06.NEPAC)

M. S. Mizuno, H. Medenwald, and D. E. P. Telenko, Department of Botany and Plant Pathology, Purdue University, West Lafayette, IN 47907-2054

SOYBEAN (*GLYCINE MAX* P29A19E)

Septoria brown spot, *Septoria glycines*
Downy mildew, *Peronospora manshurica*
White mold, *Sclerotinia sclerotiorum*

A trial was established at the Northeast Purdue Agricultural Center (NEPAC) in Whitley County, Indiana. The experiment was a randomized complete block design with six replications. Plots were 30 feet wide and 400 feet long and consisted of 24 rows, and the two center rows were used for evaluation. The previous crop was corn. Standard practices for soybean production in Indiana were followed. Soybean cultivar P29A19E was planted in 15-inch row spacing at a rate of 150,000 seeds/acre on May 24. Delaro Complete 3.82 SC 8.0 fl oz/acre was applied on August 23 at beginning seed (R5) growth stage using two different applicators: a Case IH 2250 Patriot sprayer equipped with a 30-foot boom, fitted with 18 AITTJ60-11008VP nozzles spaced 20 inches apart, at 8 mph and applied in 15 gal/acre and 60 psi and a DJI Agras T30 drone with spray pattern using 16 XRTTeeJet 11001VS nozzles at 10-foot altitude at 15 mph applied at 2 gal/acre and at 8 mph applied at 5 gal/acre. Disease ratings were assessed on September 5 at full seed (R6) growth stage. Downy mildew (DM) was rated in the upper canopy, and Septoria brown spot (SBS) was rated in the lower canopy. Disease severity of each disease was visually assessed as the percentage (0–100%) of symptomatic in three locations for each plot. White mold disease incidence was assessed by counting the number of plants in each plot with symptoms. For white mold disease severity, each plant that was observed was rated according to the following disease category: 0 = no disease, 1 = lateral branches with white mycelium and lesions, 2 = main stem with white mycelium and sclerotia present, 3 = entire plant wilted/plant death. The disease severity index (DSI) was calculated by multiplying the average number of plants in each severity category by the incidence: DSI = sum (disease severity score × number of plants)/maximum disease score × disease incidence × 100. Soybean plots were harvested on October 9, and yields were adjusted to 13% moisture. All data were analyzed in SAS 9.4 (SAS Institute, Cary, NC). A generalized linear mixed model analysis of variance was performed using PROC GLIMMIX. Values are least squares means, and values with different letters are significantly different based on Fisher's least significant difference (α = 0.05).

In 2024, weather conditions were moderately favorable for disease. White mold was the most prominent disease in the trial and reached moderate severity; sudden death syndrome was also noted in the trial. There was no significant effect between application types and nontreated control for SBS, DM, and white mold on September 5 (Table 50). There was no significant difference between treatments for yield of soybean.

TABLE 50. *Effect of Different Application Methods on Disease Severity and Yield of Soybean*

TREATMENT, APPLICATION EQUIPMENT, AND GPA[z]	SBS[y] %	DM[y] %	WHITE MOLD DI[x]	WHITE MOLD DSI[x]	YIELD[w] BU/ACRE
Nontreated control	1.5	0.7	2.6	25.7	48.6
DJI Agras T10 Drone at 2 GPA	1.1	0.6	1.3	12.8	49.6
DJI Agras T10 Drone at 5 GPA	1.1	0.7	1.7	15.7	50.3
Ground-rig at 20 GPA	1.0	0.7	1.6	15.7	48.9
P-value[v]	0.2276	0.9180	0.4843	0.4775	0.0922

[z] Fungicide applications were made on August 23 at beginning seed (R5) growth stage of Delaro Complete 3.82 SC at 8.0 fl oz/acre plus a nonionic surfactant (Preference) at a rate of 0.25% v/v. A blue tracer dye was used at 0.25% v/v. GPA = gallons per acre.

[y] Foliar disease incidence was rated on scale of 0–100% of plants with disease symptoms on September 5 at full seed (R6) growth stage. SBS = Septoria brown spot in lower canopy; DM = downy mildew in upper canopy.

[x] White mold DI = disease incidence % per plot. The white mold disease severity index (DSI) was calculated by multiplying the average number of plants in each severity category by the incidence: DSI = sum (disease severity score x number of plants)/maximum disease score x disease incidence x 100.

[w] Yields were adjusted to 13% moisture and harvested on October 9.

[v] All data were analyzed in SAS 9.4 (SAS Institute, Cary, NC). A generalized linear mixed model analysis of variance was performed using PROC GLIMMIX. Values are least squares means, and values with different letters are significantly different based on Fisher's least significant difference ($\alpha = 0.05$).

SOUTHEAST PURDUE AGRICULTURAL CENTER (SEPAC)

EVALUATION OF DRONE VERSUS GROUND FUNGICIDE APPLICATION METHODS IN CORN IN SOUTHEASTERN INDIANA, 2024 (COR24-10.SEPAC)

M. S. Mizuno and D. E. P. Telenko, Department of Botany and Plant Pathology, Purdue University, West Lafayette, IN 47907-2054

CORN (*ZEA MAYS* P1136AM)

Tar spot, *Phyllachora maydis*
Gray leaf spot, *Cercospora zeae-maydis*
Southern rust, *Puccinia polysora*

A trial was established at the Southeast Purdue Agricultural Center (SEPAC) in Jennings County, Indiana. The experiment was a randomized complete block design with four replications. Plots were 30 feet wide and 590 feet long and consisted of 12 rows, and the two center rows were used for evaluation. The previous crop was soybean. Standard practices for grain corn production in Indiana were followed. Corn hybrid P1136AM was planted in 30-inch row spacing at a rate of 32,000 seeds/acre on May 20. Veltyma 3.34 S 7.0 fl oz/acre was applied on October 9 at milk (R3) growth stage using two different applicators: an Apache AS720 sprayer equipped with a 30-foot boom, fitted with six TTJ60-11005 nozzles spaced 15 inches apart, at 12 mph and applied at 20 gal/acre and 60 psi and a DJI Agras T30 drone with spray pattern using 16 XRT TeeJet 11001VS nozzles at 12-foot altitude at 11.6 mph to apply at 2 gal/acre and at 4.7 mph to apply at 5 gal/acre. Disease ratings were assessed on September 17 at dent (R5) growth stage. Tar spot stromata severity, gray leaf spot (GLS), and southern rust were visually assessed as a percentage (0–100%) of symptomatic leaf area at ear leaf on five plants per plot at three locations in each plot and averaged before analysis. Percent canopy green was rated by visually assessing the percentage (0–100%) of canopy green on September 17 at dent (R5) growth stage. The trial was harvested on October 22, and yields were adjusted to 15% moisture. Data were averaged before analysis and subjected to mixed model analysis of variance in SAS 9.4 (SAS Institute, Cary, NC). A generalized linear mixed model analysis of variance was performed using PROC GLIMMIX. Values are least squares means, and values with different letters are significantly different based on Fisher's least significant difference (α = 0.05).

In 2024, weather conditions were not favorable for disease. Tar spot and GLS were the most prominent diseases in the trial and reached low severity. Veltyma sprayed with drone at 2 and 5 gal/acre and the ground rig significantly reduced tar spot and GLS severity over nontreated control, but there were no differences between application methods (Table 51). There were no significant differences between treatments for southern rust severity. There was no significant difference in treatments for percentage of canopy greenness, harvest moisture, and yield of corn.

TABLE 51. *Effect of Different Application Methods on Foliar Disease Severity, Canopy Greenness, and Yield of Corn*

APPLICATION EQUIPMENT AND GPA[z]	TAR SPOT[y] %	GLS[y] %	SOUTHERN RUST[y] %	CANOPY[x] GREEN %	HARVEST MOISTURE %	YIELD[w] BU/ACRE
Nontreated control	2.6 a	2.0 a	0.6	49.3	18.4	250.6
DJI Agras T30 Drone at 2 GPA	1.4 b	0.7 b	0.4	57.9	18.9	254.1
DJI Agras T30 Drone at 5 GPA	1.5 b	0.7 b	0.1	57.7	19.1	255.2
Ground-rig at 20 GPA	1.0 b	0.4 b	0.0	62.4	18.6	244.5
P-value[v]	0.0003	0.0045	0.4456	0.2473	0.5071	0.3753

[z] Fungicide treatment was applied on October 9 at milk (R3) growth stage using a ground-rig and a drone with 2 GPA and 5 GPA. All foliar treatments contained Veltyma 3.34 S at 7.0 fl oz/acre and a nonionic surfactant Maatyx at 1 oz/acre. A hot pink tracer dye was used at 0.25% v/v. GPA = gallons per acre.

[y] Foliar disease severity was visually assessed as a percentage (0–100%) of leaf area on five plants in each plot at the ear leaf on September 17 at dent (R5) growth stage. GLS = gray leaf spot.

[x] Canopy greenness was visually assessed as a percentage (0–100%) of canopy green on September 17.

[w] Yields were adjusted to 15.5% moisture and harvested on October 22.

[v] All data were analyzed in SAS 9.4 (SAS Institute, Cary, NC). A generalized linear mixed model analysis of variance was performed using PROC GLIMMIX. Values are least squares means, and values with different letters are significantly different based on Fisher's least significant difference (α = 0.05).

EVALUATION OF DRONE VERSUS GROUND FUNGICIDE APPLICATION METHODS IN SOYBEAN IN SOUTHEASTERN INDIANA, 2024 (SOY24-07.SEPAC)

M. S. Mizuno and D. E. P. Telenko, Department of Botany and Plant Pathology, Purdue University, West Lafayette, IN 47907-2054

SOYBEAN (*GLYCINE MAX* P34A98E)

Septoria brown spot, *Septoria glycines*
Downy mildew, *Peronospora manshurica*

A trial was established at the Southeast Purdue Agricultural Center (SEPAC) in Jennings County, Indiana. The experiment was a randomized complete block design with five replications. Plots were 30 feet wide and 836 feet long and consisted of 24 rows, and the two center rows were used for evaluation. The previous crop was corn. Standard practices for soybean production in Indiana were followed. Soybean cultivar P34A98E was drilled in 15-inch row spacing at a rate of 140,000 seeds/acre on April 29. Delaro Complete 3.82 SC 8.0 fl oz/acre was applied on July 24 at beginning pod (R3) and August 14 at beginning seed (R5) growth stages using two different applicators. An Apache AS720 sprayer equipped with a 30-foot boom, fitted with six TTJ60-11005 nozzles spaced 15 inches apart, was used at 12 mph and applied at 20 gal/acre and 60 psi, and a DJI Agras T30 drone with spray pattern using 16 XRTTeeJet 11001VS nozzles at 12-foot altitude at 11.6 mph was used to apply at 2 gal/acre and at 4.7 mph to apply at 5 gal/acre. Disease rating was assessed on September 14 at beginning maturity (R7) growth stage. Septoria brown spot (SBS) was rated in the lower canopy, and downy mildew (DM) was rated in the upper canopy. Severity of each disease was visually assessed as the percentage (0–100%) of symptomatic plants in three locations each plot. Soybean plots were harvested on October 25, and yields were adjusted to 13% moisture. All data were analyzed in SAS 9.4 (SAS Institute, Cary, NC). A generalized linear mixed model analysis of variance was performed using PROC GLIMMIX. Values are least squares means, and values with different letters are significantly different based on Fisher's least significant difference ($\alpha = 0.05$).

In 2024, weather conditions were not favorable for disease. SBS was the most prominent disease in the trial and reached low severity. There was no significant difference between application type and nontreated control for SBS and DM (Table 52). There was no significant difference between treatments and the nontreated control for yield of soybean.

TABLE 52. *Effect of Different Application Type on Disease Severity and Yield of Soybean*

APPLICATION EQUIPMENT, RATE, AND TIMING[z]	SBS[y] %	DM[y] %	YIELD[x] BU/ACRE
Nontreated control	1.1	0.3	58.8
DJI Agras T30 Drone 2 GPA at R3	0.9	0.1	64.9
DJI Agras T30 Drone 5 GPA at R3	0.4	0.1	61.1
Ground-rig 20 GPA at R3	0.6	0.1	58.1
DJI Agras T30 Drone 2 GPA at R5	0.7	0.2	55.3
DJI Agras T30 Drone 5 GPA at R5	0.3	0.2	64.5
Ground-rig 20 GPA at R5	0.7	0.2	57.2
P-value[w]	*0.4129*	*0.3246*	*0.5172*

[z] Fungicide treatment was applied on July 24 at beginning pod (R3) and August 14 at beginning seed (R5) growth stages using a ground rig and a drone with 2 GPA and 5 GPA. All foliar treatments contained Delaro Complete 3.82 SC 8 fl oz/acre, a nonionic surfactant of Maatyx 1 oz/acre. A hot pink tracer dye was used at 0.25% v/v. GPA = gallons per acre.

[y] Foliar disease incidence was rated on a scale of 0–100% of plants with disease symptoms on September 3 at beginning maturity (R7) growth stage. SBS = Septoria brown spot in lower canopy; DM = downy mildew in upper canopy.

[x] Yields were adjusted to 13% moisture and harvested on October 25.

[w] All data were analyzed in SAS 9.4 (SAS Institute, Cary, NC). A generalized linear mixed model analysis of variance was performed using PROC GLIMMIX. Values are least squares means, and values with different letters are significantly different based on Fisher's least significant difference (α = 0.05).

COMPARISON OF PLANTING DATES AND SEED TREATMENTS ON SOYBEAN IN SOUTHEASTERN INDIANA. (SOY24-10.SEPAC)

I. L. Miranda, J. R. Wahlman and D. E. P. Telenko, Department of Botany and Plant Pathology, Purdue University, West Lafayette, IN 47907-2054

SOYBEAN (*GLYCINE MAX* 24E453N)

Septoria brown spot, *Septoria glycines*
Downy mildew, *Peronospora manshurica*

A trial was established at Southeast Purdue Agricultural Center (SEPAC) in Butlerville, Indiana. The experiment was a randomized complete block design with three replications. Plots were 15 feet wide and 900 feet long and consisted of six rows, and the two center rows were used for evaluation. The previous crop was corn. Standard practices for soybean production in Indiana were followed. Soybean cultivar 24E453N were planted in 30-inch row spacing at a rate of 130,000 seeds/acre. Treatments were a factorial arrangement of four planting dates by four seed treatments. Soybeans were planted on April 22 (planting date 1), May 14 (planting date 2), May 30 (planting date 3), and June 12 (planting date 4). Disease ratings were assessed on August 30 at full seed/beginning maturity/full maturity (R6/R7/R8) growth stages. Septoria brown spot (SBS) and downy mildew (DM) were rated for disease severity by visually assessing the percentage of symptomatic leaf area in the upper and lower canopies. Ten roots were sampled for outer rows of each plot and rated for root rot severity on a scale of 0–100% and averaged before analysis. Each plot was harvested on October 12, and yields were adjusted to 13% moisture. All data were analyzed in SAS 9.4 (SAS Institute, Cary, NC). A generalized linear mixed model analysis of variance was performed using PROC GLIMMIX. Values are least squares means, and values with different letters are significantly different based on Fisher's least significant difference ($\alpha = 0.05$).

In 2024, weather conditions were not favorable for diseases. SBS and DM were the most prominent diseases in the trial and reached low severity. SBS severity was significantly reduced on the last planting (June 12) compared to earlier planting dates (April 22, May 14, and May 30) (Table 53). No significant differences were observed between seed treatments for SBS severity. Planting on April 22 significantly reduced DM severity compared to May 14, May 30, and June 12. The nontreated control had significantly lower DM severity compared to CruiserMaxx APX with and without Thiamethoxam and Thiamethoxam only. Planting on May 30 and June 12 significantly reduced root rot severity compared to April 22 and May 14. Moisture was significantly higher on May 14 compared to April 22, May 30, and June 12. There were no significant differences between seed treatments on root rot and moisture. Soybean yield was significantly reduced for the seeds planted on June 12 compared to May 14, which resulted on the highest yield. The seed treatments CruiserMaxx APX with and without Thiamethoxam had the highest yields compared to the nontreated control.

TABLE 53. *Effect of Planting Date and Seed Treatment on Foliar Diseases, Root Rot, and Yield of Soybean*

PLANTING DATES AND SEED TREATMENTS[z]	SBS[y] %	DM[y] %	ROOT ROT[x] %	HARVEST MOISTURE %	YIELD[w] BU/ACRE
Planting Date					
Planting date 1 (April 22)	6.1 a	2.5 c	9.6 a	12.0 b	67.5 b
Planting date 2 (May 14)	6.1 a	2.9 bc	7.5 b	12.3 a	71.8 a
Planting date 3 (May 30)	2.0 b	3.4 ab	3.6 c	12.1 b	69.7 ab
Planting date 4 (June 12)	1.0 c	3.6 a	1.8 c	12.0 b	57.0 c
Seed Treatment					
Nontreated control	3.6	2.4 b	6.2	12.1	63.9 b
CruiserMaxx APX + Thiamethoxam	3.9	3.3 a	5.4	12.2	67.3 a
Thiamethoxam	4.0	3.4 a	5.5	12.1	66.6 ab
CruiserMaxx APX without Thiamethoxam	3.6	3.3 a	5.6	12.1	68.2 a
P-value planting date[v]	0.0001	0.0090	0.0001	0.0014	0.0001
P-value seed treatment	0.1589	0.0093	0.8066	0.5017	0.0232
P-value planting date*seed treatment	0.2694	0.1524	0.0968	0.2967	0.0110

[z] Seed treatments were applied prior to planting at 10 g AI/100 kg seed.

[y] Foliar disease severity was rated on a scale of 0–100% of symptomatic leaf area in the upper and lower canopies on August 30 at full seed/beginning maturity/full maturity (R6/R7/R8) growth stages. SBS = Septoria brown spot; DM = downy mildew.

[x] Root rot was visually assessed as a percentage (0–100%) of dark discoloration on 10 roots per plot.

[w] Yields were adjusted to 13% moisture and harvested on October 12.

[v] All data were analyzed in SAS 9.4 (SAS Institute, Cary, NC). A generalized linear mixed model analysis of variance was performed using PROC GLIMMIX. Values are least squares means, and values with different letters are significantly different based on Fisher's least significant difference (α = 0.05).

APPENDIX: WEATHER DATA

TABLE 54. *Average Monthly Weather Conditions at the Purdue Agronomy Center for Research and Education (ACRE), the Pinney Purdue Agricultural Center (PPAC), the Southwest Purdue Agricultural Center (SWPAC), the Davis Purdue Agricultural Center (DPAC), the Northeast Purdue Agricultural Center (NEPAC), and the Southeast Purdue Agricultural Center (SEPAC) in Indiana, 2024[z]*

MONTHS	ACRE			PPAC			SWPAC		
	AVERAGE TEMPERATURE[y] °F	RELATIVE HUMIDITY[x] %	TOTAL PRECIPITATION[w] (INCHES)	AVERAGE TEMPERATURE[y] °F	RELATIVE HUMIDITY[x] %	TOTAL PRECIPITATION[w] (INCHES)	AVERAGE TEMPERATURE[y] °F	RELATIVE HUMIDITY[x] %	TOTAL PRECIPITATION[w] (INCHES)
January	27.3	84.1	4.69	24.9	85.1	3.33	30.4	78.7	6.01
February	38.6	72.7	0.72	35.9	74.0	0.66	43.5	63.0	0.53
March	46.9	64.4	2.35	41.8	68.9	5.13	50.8	60.5	1.88
April	54.1	69.1	5.85	50.7	68.8	4.65	59.0	66.1	8.08
May	67.3	66.6	2.14	63.0	68.4	2.86	69.5	70.9	6.12
June	73.7	64.9	2.80	71.3	66.3	2.34	76.0	64.0	2.35
July	72.7	76.5	5.01	70.2	77.7	6.16	76.1	72.3	5.62
August	71.8	76.0	4.15	69.7	76.8	2.95	75.4	71.1	2.12
September	67.0	71.0	1.42	64.9	71.1	2.39	70.7	68.2	4.98
October	57.3	61.0	0.19	55.0	63.7	9.64	60.8	60.3	0.55
November	46.2	78.6	2.74	43.8	78.1	3.37	50.0	73.6	4.26
December	33.8	79.1	3.68	31.3	76.3	2.96	38.6	74.6	5.68
Annual	*54.7*	*72.0*	*35.74*	*51.9*	*72.9*	*46.44*	*58.4*	*68.6*	*48.18*

MONTHS	DPAC			NEPAC			SEPAC		
	AVERAGE TEMPERATURE[y] °F	RELATIVE HUMIDITY[x] %	TOTAL PRECIPITATION[w] (INCHES)	AVERAGE TEMPERATURE[y] °F	RELATIVE HUMIDITY[x] %	TOTAL PRECIPITATION[w] (INCHES)	AVERAGE TEMPERATURE[y] °F	RELATIVE HUMIDITY[x] %	TOTAL PRECIPITATION[w] (INCHES)
January	27.1	83.0	3.57	26.1	85.1	4.46	30.5	79.9	7.67
February	37.3	72.0	0.70	36.5	73.1	0.90	42.3	64.6	1.30
March	45.0	67.0	3.85	43.6	67.8	3.52	48.7	63.7	3.70
April	54.0	69.0	6.29	52.3	67.4	7.96	57.5	67.4	4.95
May	66.5	68.0	1.78	65.3	67.4	3.61	67.6	74.3	4.42
June	72.7	65.0	2.43	72.8	63.5	3.06	73.4	67.7	1.23
July	71.9	75.0	2.22	72.2	75.2	4.10	74.5	74.6	5.57
August	71.1	75.0	3.83	71.3	75.1	1.67	73.1	75.1	4.82
September	66.3	70.0	2.20	67.0	70.4	1.43	68.4	72.5	4.71
October	56.0	64.0	0.20	56.4	61.6	0.42	58.0	67.0	0.66
November	45.6	78.0	3.41	45.1	79.2	1.89	49.0	76.2	5.40
December	33.7	79.0	4.27	32.8	79.7	4.46	37.9	75.5	5.05
Annual	*53.9*	*72.1*	*34.75*	*53.5*	*72.1*	*37.48*	*56.7*	*71.5*	*49.48*

[z] Data courtesy of Beth Hall, Jonathan Weaver, and Austin Pearson, Indiana State Climate Office, https://ag.purdue.edu/indiana-state-climate/. Taken from Purdue Mesonet stations.

[y] Average temperature for each month.

[x] Average relative humidity for each month.

[w] Total precipitation for each month.

ABOUT THE AUTHORS

DARCY E. P. TELENKO is an associate professor and Extension plant pathologist in the Department of Botany and Plant Pathology at Purdue University. Her interdisciplinary research and Extension program are involved in studying the biology and management of soilborne and foliar pathogens of agronomic crops. Telenko is a native of western New York and received her PhD at North Carolina State University. She has published more than sixty peer-reviewed manuscripts and two hundred Extension publications. She was awarded the 2024 Leadership Award from the Purdue University Cooperative Extension Specialist Association.

SUJOUNG SHIM is a research associate in the Department of Botany and Plant Pathology at Purdue University. Her research involves designing, conducting, analyzing, and reporting on a variety of research projects. She has a BS in pharmaceutical science and an MS in public health, both from Purdue University. Shim has served as a coauthor on more than ten peer-reviewed publications and twenty-five peer-reviewed technical reports.

www.ingramcontent.com/pod-product-compliance
Lightning Source LLC
Chambersburg PA
CBHW081133170526
45165CB00008B/2662